Physical Chemistry for the Biomedical Sciences

Physical Chemistry for the Biomedical Sciences

S.R. LOGAN
Reader in Chemistry
University of Ulster

TAYLOR & FRANCIS
ALERE FLAMMAM
1798 – 1998

UK   Taylor & Francis Ltd, 1 Gunpowder Square, London EC4A 3DE
USA Taylor & Francis Inc., 1900 Frost Road, Suite 101, Bristol, PA 19007-1598

British Library Cataloguing-in-Publication Data

A catalogue record for this book is available from the British Library.
ISBN 0–7484–0709–X (cased)
ISBN 0–7484–0710–3 (Paperback)

Library of Congress Cataloging-in-Publication Data are available

Cover design by Jim Wilkie
Typeset in Times 10/12pt by Keyword Typesetting Services Ltd
Printed by T.J. International Ltd, Padstow, UK

Contents

Contents

Contents

Preface

In the 1990s, I found myself teaching physical chemistry every autumn to a first year class containing students of Biomedical Sciences (called the *Biomeds*) and Human Nutrition (the *Hum Nuts*). To my considerable surprise, I could find no textbook that was well suited to the needs of this group. After some deliberation, I resolved to offer my own services towards writing a book of this nature. This is the result.

Students entering degree courses in the field of Biomedical Sciences come, in my experience, from a range of educational backgrounds. Most have studied biology, but whereas some are also well versed in the 'hard' sciences, others may have opted for other school subjects, only to decide later that they wanted a biomedical career. No book can be ideally geared to all parts of such a broad spectrum, but it was deemed desirable to proceed, as far as possible, from first principles rather than from facts that might be familiar only to those who had scored well in A level Chemistry.

Although these students are likely to become graduates who utilise, rather than develop, physical chemistry, some minimum mathematical content is essential for them. The days of a *Grundriss der physikalische Chemie* containing not a single equation are surely over. But even if physical chemistry is to be made no more mathematical than it has to be, one must include logarithms and calculus, since these are necessary to express the relationships of nature in a succinct way. So these topics, which are prime examples of scientifically useful maths that is not in the curriculum of GCSE Mathematics, are briefly introduced in the Appendices. However, this can scarcely make good the total deficit in mathematical preparation, for recent 'reforms' in the teaching of maths have also diminished the amount of algebra done in schools, to the extent that the notion of there being a solution to a mathematical equation is no longer as familiar as in earlier times.

Curriculum changes and other factors have also conspired to produce university entrants with a weaker background in physics. To take account of this, another Appendix has been added for handy reference, and the unfamiliarity of many potential users with concepts such as work, heat, waves and electrostatic interactions has been borne in mind when introducing these topics in the text.

My selection of topics within physical and general chemistry is derived from my perception of those areas that are required as a background for studying the various roles of molecules within biological systems. In some places, there is arguably rather more theory than is strictly necessary. In part, I have done this for the benefit of the keen student who wants to know that little bit more, or where this little bit more may make the subject easier to appreciate.

In presenting chemical bonding, I have opted to treat covalency only in terms of the molecular orbital (MO) theory, though I do try to make clear that this represents only one of the possible approximations that are necessary in systems that are more complex than the hydrogen molecule ion. My reasons for so doing are threefold. It is surely less confusing for students to be presented with just one system than to have to contend with two different and mutually inconsistent versions. Also, MO theory can act as a useful basis for interpreting electronic transitions and it is widely used to interpret the mechanisms of organic reactions.

Elsewhere, it may be thought unusual that I mention the non-ideality of gases without bringing in the Van der Waals equation. In this context, I see the empirical approach of the virial equation as having more merit. Moreover, the equation of Van der Waals does not readily lead to the sort of potential function that applies between real molecules. Also, I may be thought negligent in delaying the introduction of the activity coefficient as long as I do. My message is that behaviour deviates appreciably from ideality only with electrolytic solutions, and a major manifestation of this arises in regard to electrochemical cells. However, I must admit that in some of the ionic equilibria considered in Chapter 9, the deviation of the activities from the molar concentrations is not negligible. My case is that by dwelling more on this matter, I might well have been obscuring other difficulties that students have with chemical equilibria among ions.

Chemical kinetics has been left to the last, as befits that branch of physical chemistry which, it can be argued, is intellectually the most difficult. No other branch possesses an entire empirical framework, though most have the occasional empirical quantity, like the decadic absorption coefficient or the second virial coefficient or the activity coefficient, necessary to document actual as distinct from ideal behaviour. The essence of the difficulty is the interrelationship of these empirical parameters to those quantities that emerge from a theoretical development of the reaction, whether one assumes that it is a single step or a multi-step reaction. To keep things simple, Chapter 11 deals with experimental matters and the empirical approach to documenting what is observed, leaving this interplay of theory and experiment to Chapter 12.

For the most part, SI units have been used because of their innate self-consistency. Thus, volumes are given in cm^3 or dm^3 rather than as ml or litres (l). However, in areas where other units are widely used, such have been introduced and the relationship to the SI unit is made clear. For concentrations of solutions, I use the style mol dm^{-3}, as favoured by the Royal Society of Chemistry, rather than mol l^{-1} or M.

Over the years, I have put many numerical and other logical problems in front of my classes of biomeds and hum nuts. Working through these problems has helped the students get to grips with the concepts and the equations that my lectures have introduced. Typically this involves pocket calculator calculations and occasionally it requires the plotting of a graph. So each chapter closes with a selection of problems, in the hope that these can benefit a wider audience.

In preparing this text, I have had the benefit of the observations of Bill Byers, my inorganic colleague on the Jordanstown campus, on the opening chapters and those of my publisher's reviewer on the entire book. Additionally, Iona Hamilton, a former participant on my course and now a postgraduate in the 5*-rated School of Biomedical Sciences, looked at the book from a student's viewpoint. Yvonne Barnett, a colleague in the same School, forwarded helpful comments on Chapter 10 and organised an instructive analysis of Chapters 11 and 12 by Birgitte Munch-Petersen of Roskilde, Denmark. David Thurnham, also in the School of Biomedical Sciences, gave me useful comments on the later sections of Chapter 12. I also acquired several useful ideas from other colleagues in Coleraine. Thomas Miller ran lots of spectra for me. The typing was done partly by Margaret Avenell in the School office, and principally by my wife, Renée. Bill Byers also helped check the proofs. I thank them all for their willing and skilled assistance, and Elaine Stott and Jackie Day of the publishers for their enthusiastic support.

S.R. Logan
University of Ulster at Coleraine
October, 1997

1

The atom and the atomic nucleus

This introductory chapter looks at the historical development of the concept of the atom and the discovery of its basic make-up. The behaviour of the atomic nucleus is outlined in terms of the classical nuclear components, the proton and the neutron.

Nuclide stability is shown to be determined principally by two issues: is the nucleus too large and has it an acceptable ratio of neutrons to protons? Thus certain nuclides are stable while others undergo disintegration processes that are readily explicable in terms of the above criteria, and lead to the emission of various radiations. In all cases, radioactive decay of a nuclide is shown to follow an equation for exponential decrease.

1.1 Introduction

Among the philosophers of ancient Greece, there were those who argued that matter could be subdivided indefinitely and those who held the contrary view. According to the latter school, matter was composed of discrete particles, for which the name *atom* was coined, meaning that which could not be cut.

On this issue of the physical nature of matter, essentially no progress was made until the early nineteenth century. In his book, published in 1808, Dalton postulated not merely that atoms existed but that the properties of atoms were peculiar to the element involved. Thus, he asserted, every atom of oxygen was identical, as was every atom of carbon, but these two types of atom were distinct. This is the kernel of Dalton's Atomic Theory and the key to understanding chemistry. Only the term 'atom' had been supplied from elsewhere. The ideas were all produced by Dalton who, ironically but significantly, was much more a general scientist than a chemist.

At that stage, chemists were already using the terms **element** and **compound** on the basis of behaviour. If a substance could be broken down into other substances then it could not possibly be an element, but it could, if pure and

1

homogeneous, be a compound. Also, a substance produced by the combination of two elements, like sulfur and oxygen, was a compound. Dalton's proposals allowed all of this to be explained in terms of composition. He proposed that compounds are formed when atoms of different elements combine together in simple ratios. Thus, for example, it could be conceived that carbon and oxygen might combine in the ratio of 1:1 or in the ratio of 1:2, leading to the compounds carbon monoxide, CO, and carbon dioxide, CO_2.

While the postulates of Dalton proved not to be absolutely correct in every respect, they served as a framework on the basis of which chemical reactions and the structures of compounds could be understood. Their importance is clearly attested by the phenomenal progress that was achieved, particularly in the sphere of organic chemistry, during the nineteenth century. However, it is interesting to note that all of this was done without obtaining an answer to one of the most elementary of questions: how big is an atom? Dalton did not know, and, wisely, did not commit himself on a matter on which he had no information. Others realised that their lack of knowledge on this point need not inhibit their making progress within chemistry. No reliable idea of the size of an atom was available until around 1909, by which time chemistry had made quite considerable progress.

1.2 The structure of the atom

In the light of Dalton's Atomic Theory, an atom was a small particle whose chemical combination with other atoms could be understood on the working assumption that it had a certain number of tentacles attached to it, capable of linking with the tentacles of other atoms. There remained the question of the structure and composition of an atom.

Experiments on the conduction of electricity through gases led to the realisation that particles, called **cathode rays**, were emitted from the negatively charged electrode. These rays were found to be deflected by the application of an electric field or of a magnetic field. Thomson (JJ) concluded that 'they are charges of negative electricity carried by particles of matter', and made a good estimate of their charge-to-mass ratio. It could only be assumed that this was a subatomic particle or, more explicitly, that it was a component of an atom, and it was named the **electron**.

The most influential experiment in regard to the discovery of the structure of the atom was undoubtedly that of Rutherford in 1911. This involved the use of the α-particle, which is introduced in section 1.5 and which was known at that time to be positively charged. A beam of α-particles was allowed, in a vacuum, to impinge on a thin metal foil. This was found to cause deflection of the α-particles, to variable extents. Many were unaffected by the presence of the foil, some were deflected by small angles, but a few were deflected by very large angles, approaching $180°$.

In attempting to explain this scattering pattern, Rutherford realised that, since so many α-particles could pass through unaffected, the large deviations suffered by some could not arise as the consequence of multiple small deflections. The only way he could explain his observations was by assuming that the target material consisted of atoms in which substantial positive charges were concentrated in very small volumes. If the α-particle were directed straight towards this positive charge, it would be repelled; if it were directed on a path close to this positive charge, the α-particle would follow a parabolic path whose curvature would depend on the exact geometry of approach; and if an α-particle did not come close to any centre of atomic positive charge, the resulting deflection might be minimal. Rutherford showed that his model would predict quantitative agreement with his results.

So, an atom was a highly non-uniform entity, with a large positive charge concentrated in a tiny **nucleus**, in which resided the bulk of the mass of the atom. The remainder of the space was presumably occupied by that number of electrons required to preserve overall electrical neutrality.

After the Rutherford experiment which led to the discovery of the atomic nucleus, the subatomic particles of which it is composed were detected within twenty years. There are two of these, called the **proton** and the **neutron**. The proton has a positive charge equal (and opposite) to that of the electron and a mass approximately 1840 times greater. The neutron is just a little heavier than the proton but carries no charge and is thus more difficult to detect. Chadwick's discovery of the neutron in 1932 completed the set of nuclear components, but did not end the endeavours of nuclear physicists to investigate the secrets of the atom. From a chemist's viewpoint, the structure of the atom can adequately be described on the basis of the state of knowledge existing in 1935.

1.3 The atomic nucleus

If we treat protons and neutrons as nuclear components, let us suppose that a nucleus consists of Z of the former and N of the latter. It might be assumed that any combination of Z and N is possible. In practice, only a small fraction of these represent stable nuclei, as is demonstrated in Figure 1.1. From this, the following may be seen:

1 Every nucleus containing more than one proton contains at least one neutron.

2 For Z values up to around 20, the preferred ratio of N to Z is around 1:1, but at higher values of Z, the preferred ratio rises towards 1.5:1.

3 There exist no stable species of nuclei (called **nuclides**) with Z greater than 83.

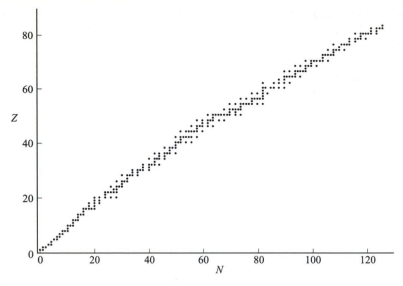

Figure 1.1   A representation of the stable nuclides in terms of the combinations of the values of Z against N, using the data quoted in Friedlander, G., Kennedy J. W., Macias E. S. and Miller J. M., 1981, *Nuclear and Radiochemistry*, 3rd Edn, Wiley: New York, pp. 610-50.

4   Whereas for some values of Z there are many stable nuclides, for others, such as $Z = 43$ or 61, there exist none. Likewise, for some values of N there are many stable nuclides, but for others, such as $N = 19$, 21, 35 and 39, there are none.

The model of the atomic nucleus which we are employing can qualitatively explain some of these phenomena. In a nucleus with Z greater than 1, there must inevitably be the problem of the mutual repulsions arising out of the positive charges. It would seem that this cannot be overcome unless neutron components are present, and this role will tend to become more important with increasing values of Z.

One fundamental fact about atomic nuclei is that the nucleus weighs less than the sum of the weights of its components. The magnitude of this discrepancy is always less than 1%, but it is of immense significance. Einstein's Theory of Relativity led to the view that mass and energy were interconvertible. On that basis, put crudely, the loss of mass is a measure of the binding energy within the nucleus, in accordance with the Einstein formula,

$$E = \Delta m.c^2 \tag{1.1}$$

which expresses the energy E in terms of the mass difference, Δm and c, the velocity of light.

It is convenient to introduce the parameter A, equal to the sum of Z and N, and to call it the number of **nucleons**, or nuclear components, within a nucleus.

To represent graphically the variations in the nuclear binding energy, the natural parameter to use as the abscissa is A, the number of nucleons. Additionally, the plot becomes more informative if the figure plotted against A is the binding energy per nucleon, and this is shown in Figure 1.2. This diagram shows that as A increases, the binding energy per nucleon steadily rises, attaining a maximum value at around $A = 60$. After this, the binding energy per nucleon declines until at $A = 209$, it is about 10% less than at its peak. The fact and the manner of this decrease can be shown to be sufficient reason for point 3, highlighted above.

1.4 Nuclides and chemistry

Of the three parameters that have been used in regard to particular species of atom, namely Z, N and A, it can reasonably be claimed that the first is the most important. In a neutral atom, the nucleus is surrounded by electrons of a number equal to Z and, as we shall see, the chemical behaviour of the atom is determined by the number of these electrons, called the atomic number. Thus

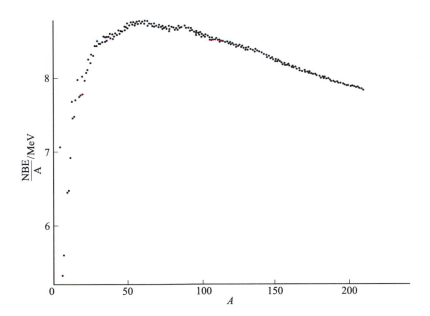

Figure 1.2   A plot of the nuclear binding energy per nucleon, NBE/A, against the mass number, A, for some of the stable nuclides, using data quoted in Friedlander, G., Kennedy, J. W., Macias E. S. and Miller, J. M., 1981, *Nuclear and Radiochemistry*, 3rd Edn, Wiley: New York, pp. 610–50.

each Z value corresponds to a single chemical element, starting with 1 for hydrogen and 2 for helium. For each element, there is a symbol, consisting of either one or two letters, of which the first is always written as a capital and the second (where present) is always lower case. For the above elements, the symbols H and He are used.

The above three parameters lead on to three terms applicable to nuclides. Those with the same value of the atomic number, Z, are called **isotopes**. Where the number of neutrons, N, is the same, they are **isotones** and if the number of nucleons, A, is the same, they are **isobars**. The first of these terms is the most important.

In order to specify a particular stable nuclide, of which there are 263, it is not sufficient to quote the value of Z or the name of the element. Since stable nuclides exist with only 81 different values of Z, many elements have several stable isotopes. The system widely used is to write the A value as a superscript and the Z value as a subscript before the symbol of the element, as shown below:

$$\ce{^1_1H} \qquad \ce{^4_2He} \qquad \ce{^{13}_6C} \qquad \ce{^{209}_{83}Bi}$$

In a sense the Z value is redundant, but its inclusion can be very helpful, especially for the later elements whose atomic numbers may not be uppermost in one's mind.

It is interesting to look at the Z and N values of the 263 stable nuclides. A simple categorisation of each of these as being either even (i.e. exactly divisible by two) or odd, leads to the following subtotals:

Even Z, even N	156
Even Z, odd N	52
Odd Z, even N	50
Odd Z, odd N	5
	263

Our simple model of the atomic nucleus affords no explanation as to why even values should be so much more stable than odd values, which is highlighted by the fact that the stable odd–odd nuclides are $\ce{^2_1H}$, $\ce{^6_3Li}$, $\ce{^{10}_5B}$, $\ce{^{14}_7N}$ and $\ce{^{50}_{23}V}$.

The consequence is that there are 208 nuclides with even atomic number, Z, and only 55 with odd atomic numbers. Not surprisingly, the elements with a larger number of stable isotopes all have even atomic numbers. For example, tin ($\ce{_{50}Sn}$) has 10, with A values equal to 112, 114, 115, 116, 117, 118, 119, 120, 122 and 124. Elements with an odd atomic number may have very few stable isotopes. Technetium ($\ce{_{43}Tc}$) has none and aluminium has only one, $\ce{^{27}_{13}Al}$; but several, such as chlorine, have two: $\ce{^{37}_{17}Cl}$ and $\ce{^{35}_{17}}$. Also, the natural abundances of elements with even atomic numbers tend to be rather greater than are those of odd atomic number.

1.5   Unstable nuclides and radioactivity

In 1896, Becquerel discovered that photographic plates in his laboratory were being fogged, even though they had not been exposed to light. Investigation revealed that the cause was the proximity of the covered plates to certain materials in the laboratory, and that the fogging was caused by penetrating rays which these materials emitted. This was the discovery of radioactivity, which is the consequence of nuclide instability. The major modes of nuclide disintegration may be classed under three headings.

1.5.1   α-Particle emission

An α-particle is the nucleus of the atom $^4_2$He, and among small nuclides it has an anomalously high nuclear binding energy. For the larger nuclides, with Z greater than 83, a quite common disintegration process is the emission of an α-particle, yielding a product nucleus in which both Z and N are smaller by 2. For example:

$$^{238}_{92}\text{U} \quad \rightarrow \quad ^{234}_{90}\text{Th} \quad + \quad ^{4}_{2}\alpha \tag{1.2}$$

In a nuclear process of this type, there is invariably a balance between the aggregate values of the numbers of nucleons before and after. However, there is not usually a mass balance and unless the total mass of the products is *less* than that of the original nucleus, the process will not occur. This means, in effect, that the total nuclear binding energy of the products must be greater than that of the original nuclide. For process (1.2), a decrease in mass does indeed occur, for reasons referred to in section 1.3. In accordance with the Einstein formula, this is converted into energy and goes into kinetic energy of the products. Consequently, the α-particles produced in nuclear reaction (1.2) have a kinetic energy of 4.2 MeV.

1.5.2   β⁻-Particle emission

Figure 1.1 demonstrates that, if a nuclide is to be stable, it must have an acceptable ratio of neutrons to protons. Below and to the right of the curving line of dots is thus a region where there is an excess of neutrons, whereas above and to the left of this line represents a condition of neutron deficiency.

An example of a nuclide with excess neutrons is $^{24}_{11}$Na . This may undergo a dismutation process in which a negatively charged and very light particle, called a β^--particle, is emitted from the nucleus. The process may be summarised as follows:

$$^{24}_{11}\text{Na} \quad \rightarrow \quad ^{24}_{12}\text{Mg} \quad + \quad ^{\ 0}_{-1}\beta \tag{1.3}$$

The β^--particle is known to be the same as an electron, but when this species is generated in this way, it is normally referred to as a β^--particle. As a consequence of nuclear reaction (1.3), Z has increased by 1 and N has decreased by 1, so that the neutron excess problem has been alleviated. One way of representing reaction (1.3) is that, in effect, a neutron has been converted into a proton plus an electron.

Once again, a process like reaction (1.3) can occur only if the aggregate mass of the product species is less than that of the initial nuclide. However, the β^--particles are emitted with a range of energies, though the maximum value of this range is in agreement with the energy equivalence of the decrease in mass, using the Einstein formula. Since β^--particles are far lighter and therefore faster than α-particles, they have much greater penetration than α-particles of comparable kinetic energy.

1.5.3 β^+-Particle emission

The converse of a nuclide with excess neutrons is one which is neutron deficient, of which an example is $^{22}_{11}\text{Na}$. This may emit a β^+-particle, otherwise called a positron, which has the same mass as an electron:

$$^{22}_{11}\text{Na} \quad \rightarrow \quad ^{22}_{10}\text{Ne} \quad + \quad ^{\ 0}_{+1}\beta \tag{1.4}$$

However, the positron is not stable in the way that an electron is, and a positron plus an electron will undergo mutual annihilation to give two quanta of electromagnetic radiation.

An alternative process which a neutron-deficient nuclide may undergo is that the nucleus may capture one of the surrounding electrons:

$$^{55}_{26}\text{Fe} \quad + \quad ^{\ 0}_{-1}\text{e} \quad \rightarrow \quad ^{55}_{25}\text{Mn} \tag{1.5}$$

This leads to a similar change in the nucleus. The consequence of the nuclear processes (1.4) and (1.5) is that Z has decreased by 1 and N has increased by 1, which signifies a movement on the chart of Z against N opposite to that of process 2 in section 1.3. Both represent change of the neutron to proton ratio towards the value which is more stable.

For completeness, it should be added that in many cases the emission of α-, β^-- or β^+-particles is accompanied by the emission of a quantum of electromagnetic radiation. Where this happens, the energy equivalent of the mass decrease will be shared between the kinetic energy of the emitted particles and the energy E of the quantum. The latter is given by Planck's formula,

$$E = h\nu \tag{1.6}$$

where h represents Planck's constant and ν is the frequency of the radiation. Since the energy changes within the atomic nucleus are large, the quanta emitted in this way have high energies and are called γ-rays. They are much more penetrating than β^--particles.

For any nuclear disintegration, the rate is proportional to the number of nuclei present and this means that the number of atoms remaining, n, falls off exponentially, in accordance with the equation:

$$n = n_o \exp(-\lambda t) \tag{1.7}$$

where n_0 was the number present at time t equal to zero, and λ is a constant for that particular disintegration process.

Two consequences of this exponential relationship may be demonstrated. Firstly, taking logarithms of each side of equation (1.7) we have,

$$\ln n = \ln n_o - \lambda t \tag{1.8}$$

which means that the logarithm of the number present is a linear function of the time. Secondly, if we then put n equal to $n_0/2$, we find that the time, $t_{\frac{1}{2}}$, required for half the nuclei to disintegrate is given by,

$$t_{\frac{1}{2}} = \frac{\ln 2}{\lambda} \tag{1.9}$$

and is independent of n_o, the initial number. So for any size of sample, half will disintegrate in $t_{\frac{1}{2}}$, and half of what remains in a further period of $t_{\frac{1}{2}}$.

It is normal to denote the rapidity with which a nuclide decays by the value of the half-life, $t_{\frac{1}{2}}$. For example, for $^{210}_{84}$Po it is 138.4 days, for $^{232}_{90}$Th it is 1.4×10^{10} years, but for $^{21}_{11}$Na it is 22 seconds.

1.6 The occurrence of unstable nuclides

Some unstable (i.e. radioactive) nuclides occur naturally. The reason, in most cases, is that the half-life of the radionuclide is very long and is within an order of magnitude or so of the age of the Earth. (Note that after 10 half-lives, only 0.1% of the original amount remains, and after 20 half-lives, only 10^{-4}%.) One such example is $^{238}_{92}$U for which the half-life is 4.5×10^9 years. There are no stable isotopes of uranium: this is the most abundant one, and the one with the longest half-life. Another example is $^{87}_{37}$Rb with a half-life of 4.7×10^{10} years, which is the less abundant and an unstable isotope of rubidium.

In addition to these radionuclides and their decay products, others exist naturally with appreciably lower half-lives. An example is $^{14}_6$C, with a half-life of 5770 years. The presence of this species is attributable to its continual generation in the upper atmosphere by cosmic ray action on atoms of $^{14}_7$N:

$$^{14}_7\text{N} \;+\; ^{1}_0\text{n} \;\rightarrow\; ^{14}_6\text{C} \;+\; ^{1}_1\text{p} \tag{1.10}$$

Artificial radionuclides may be generated by bombarding an appropriate sample with subatomic particles, such as neutrons, protons or deuterons. For example, ^{59}Co, if placed in an atomic pile, undergoes a neutron capture process,

$$^{59}_{27}Co \; + \; ^{1}_{0}n \; \rightarrow \; ^{60}_{27}Co \; + \; \gamma \tag{1.11}$$

leading to the formation of the radionuclide ^{60}Co.

The total number of known artificial radionuclides is in excess of 1500, and includes all elements up to atomic number Z equal to 115. No stable nuclides have been discovered in addition to those which occur naturally.

1.7 Atomic mass units and the mole

In order to denote the masses of various nuclides, atoms and subatomic particles it is useful to have a convenient numerical scale. The first instinct was to put the mass of the hydrogen atom equal to one, but this proved to be a poor choice. The scale currently used is based on the mass of the carbon-12 atom, $^{12}_{6}C$, being precisely 12.

On this basis the mass of a proton is 1.007277 and that of an electron is 0.0005486. Adding these together we have 1.007825 for the mass of the hydrogen atom, $^{1}_{1}H$, and it is convenient to use this figure in calculating the masses of the components of an atom, since there will be the same number of protons incorporated in the nucleus as there are electrons around it. On the same scale, known as atomic mass units (a.m.u.), the mass of the neutron is 1.008665. We may use these to evaluate the mass of the components of the sodium-23 atom, $^{23}_{11}Na$.

$$\begin{aligned} \text{Mass of components} \; &= 1.007825Z + 1.008665N \\ &= 11 \times 1.007825 + 12 \times 1.008665 \\ &= 23.19006 \end{aligned} \tag{1.12}$$

It is important to distinguish some interrelated but distinct quantities. The number of nucleons, A, also called the mass number, is necessarily an integer. As illustrated above, the aggregate mass of the components of an atom cannot be an integer. The actual mass is always less than this (except in the case of $^{1}_{1}H$): for sodium-23 it is 22.98977, which is less by 0.86%.

If there is only one naturally occurring isotope of the element, as there is in the case of sodium, then this actual mass of the atom, on the a.m.u. scale, is also the atomic mass of the element. However, where there is more than one isotope present, the atomic mass is a weighted mean of the actual masses of each isotope, with weighting factors proportional to the respective natural abundances. For example, chlorine has two stable isotopes, $^{35}_{17}Cl$ with an abundance of 75.53% and ^{37}Cl with 24.47%. So the atomic mass of chlorine may be calculated as follows from the actual masses of these atoms:

$$\begin{aligned} \text{Atomic mass of Cl} \; &= \frac{(75.53 \times 34.96885) + (24.47 \times 36.96590)}{100} \\ &= 35.458 \end{aligned} \tag{1.13}$$

A list of the names of the first 56 elements, along with the symbol, atomic number and atomic mass is given in Table 1.1.

To relate the atomic mass scale to the ordinary SI unit of mass, Avogadro's constant is defined as the number of atoms of $^{12}_{6}C$ in 0.012 kg (= 12 g) of carbon-12. This number, with a value of 6.0221×10^{23}, lies in a range which is far beyond all human experience, and means that the radius of an atom is of the order of 10^{-10} m. Avogadro's constant of any entity is defined as one **mole** of that species. So 1 mole of $^{12}_{6}C$ atoms weighs 12 g precisely, and 1 mole of chlorine atoms (with the natural abundance of ^{35}Cl and ^{37}Cl) weighs 35.457 g. As we shall see, elemental chlorine exists as diatomic molecules, Cl_2, so 1 mole of elemental chlorine weighs 70.914 g.

We are now in a position to quantify the energy changes which accompany nuclide disintegration processes. Let us consider the process:

$$^{212}_{84}Po \quad \rightarrow \quad ^{208}_{82}Pb \quad + \quad ^{4}_{2}\alpha \tag{1.14}$$

The decrease in mass, Δm, is in a.m.u. equal to $211.9886 - (207.97664 + 4.00260) = 0.00962$. Dividing this by $10^3 N_A$, where N_A is Avogadro's constant, converts this to kg. So, for the energy release accompanying process (1.14) we have, using equation (1.1),

$$E = \Delta m.c^2$$
$$= \left(\frac{0.00962}{6.022 \times 10^{26}} kg \right) \times (2.998 \times 10^8 \, m \, s^{-2})^2$$
$$= 14.36 \times 10^{-13} \, J \tag{1.15}$$

For events on the atomic scale, the Joule (J) is rather a large unit of energy. A more fitting unit is the **electron volt** (eV), which is equal to $1.602 \times 10^{-19} J$, so that the energy release for process (1.14) is 8.96×10^6 eV (or 8.96 MeV).

As we shall see, when chemical reactions take place, the energy changes involved, per atom or per molecule involved, are on the scale of a few electron volts. The above calculation suffices to illustrate that for reactions involving atomic nuclei, the energy released is around a million times greater. So conventional power stations consume thousands of tons of coal or oil, whereas nuclear power stations work with very modest amounts of uranium or plutonium.

1.8 The use of isotopes as labelled atoms

In science and in medicine, the existence of more than one isotope of an element can provide a tool for further investigation. In one sense, these applications fall into two categories: those using a stable isotope and those based on a radioactive isotope. In each case, the basic supposition is that the labelled atoms behave in just the same way as the other atoms of the same element.

11

Table 1.1   Details of the first 56 elements

Atomic number	Name	Symbol	Atomic mass
1	Hydrogen	H	1.008
2	Helium	He	4.003
3	Lithium	Li	6.939
4	Beryllium	Be	9.012
5	Boron	B	10.81
6	Carbon	C	12.01
7	Nitrogen	N	14.01
8	Oxygen	O	16.00
9	Fluorine	F	19.00
10	Neon	Ne	20.18
11	Sodium	Na	22.99
12	Magnesium	Mg	24.31
13	Aluminium	Al	26.98
14	Silicon	Si	28.09
15	Phosphorus	P	30.97
16	Sulfur	S	32.06
17	Chlorine	Cl	35.45
18	Argon	Ar	39.95
19	Potassium	K	39.10
20	Calcium	Ca	40.08
21	Scandium	Sc	44.96
22	Titanium	Ti	47.90
23	Vanadium	V	50.94
24	Chromium	Cr	52.00
25	Manganese	Mn	54.94
26	Iron	Fe	55.85
27	Cobalt	Co	58.93
28	Nickel	Ni	58.71
29	Copper	Cu	63.54
30	Zinc	Zn	65.37
31	Gallium	Ga	69.72
32	Germanium	Ge	72.59
33	Arsenic	As	74.92
34	Selenium	Se	78.96
35	Bromine	Br	79.91
36	Krypton	Kr	83.80
37	Rubidium	Rb	85.47
38	Strontium	Sr	87.62
39	Yttrium	Y	88.91
40	Zirconium	Zr	91.22
41	Niobium	Nb	92.91
42	Molybdenum	Mo	95.94
43	Technetium	Tc	[99.0]
44	Ruthenium	Ru	101.07
45	Rhodium	Rh	102.91

Table 1.1 (*cont*)

Atomic number	Name	Symbol	Atomic mass
46	Palladium	Pd	106.4
47	Silver	Ag	107.87
48	Cadmium	Cd	112.40
49	Indium	In	114.82
50	Tin	Sn	118.69
51	Antimony	Sb	121.75
52	Tellurium	Te	127.60
53	Iodine	I	126.90
54	Xenon	Xe	131.30
55	Caesium	Cs	132.91
56	Barium	Ba	137.34

To detect the abundance of a stable isotope, such as $^{13}_{6}C$ or $^{2}_{1}H$ (frequently called deuterium and denoted by D), the most widely used method is mass spectrometry. This involves the ionisation of the molecules to generate positively charged ions, which are then separated on the basis of their mass-to-charge ratio. Both the isotopic examples cited above are heavier, by 1 mass unit, than the 'normal' isotope, so if one atom of either is present in a molecule, the mass of the resulting ion will be greater by 1 a.m.u. Both ^{13}C and D have finite natural abundances, but when these isotopes are used as tracers, it is at vastly higher concentrations.

An example of a biomedical use of a stable isotope would be feeding to a subject a hexose (a sugar molecule containing six atoms of carbon) in which one of these six atoms is substituted with ^{13}C. The fate of the hexose within the subject can be monitored by collecting the carbon dioxide which is breathed out and by analysing it, using a mass spectrometer, for its ^{13}C content.

Radionuclides emit ionising radiations which are potentially damaging, so their use inevitably raises issues of safety. However, these ionising radiations can permit the radionuclides to be detected with high sensitivity, so that they can be used in very low abundances.

For experiments *in vivo*, where safety aspects are paramount, it is highly desirable that there exists a suitable radionuclide with a half-life of the order of a few days. The shorter the half-life, the smaller is the amount of radioactive substance required to achieve the necessary number of disintegrations per second, and thus, the smaller is the total amount of radiation liable to be absorbed by the subject. The minimum half-life that can be contemplated will be a function of the time required to fetch the radionuclide from the supplier and conduct the required experiment.

Studies in which radionuclides are assayed by appropriate radiation detectors produce data indicating the location of these radionuclides, but in no way reflecting their chemical state. Since γ-rays have greater penetration than α- or

β^--particles, nuclides which emit them are the easiest to detect. Also, the fraction of the energy absorbed by the medium immediately around the unstable atom is least for a γ-ray emitter.

Bone contains large amounts of the elements calcium and phosphorus, among many others. To study the structure and function of a patient's bones, it would be sensible to inject radioactive calcium or phosphorus. However, the leading candidates, ^{45}Ca and ^{32}P, are pure β^--emitters, whose radiations would be difficult to detect externally. A fairly energetic γ-ray, which could be detected externally, is emitted by $^{85}_{38}Sr$. Strontium shows chemical resemblances to calcium, so this radionuclide could perhaps serve the purpose. One problem is that ^{85}Sr has a half-life of 64 days, which means that its use entails giving the patient an undesirably high dose of radiation.

An ingenious solution to the problem is to use a form of an isotope of technetium, element 43, of which no stable isotopes exist. ^{99m}Tc is a γ-emitter and has a half-life of 6 hours, so it is easily detected and an amount which is sufficient for that purpose does not entail giving the patient a large irradiation dose. However, to be of service, the atoms of Tc have to be injected in a chemical form that will cause them to be incorporated into the bones: technetium is a transition metal and these are not usually to be found there. This objective is achieved by attaching the atoms of ^{99m}Tc to phosphorus compounds which chemically resemble the pyrophosphates that occur naturally in the body and which participate in bone formation. In this way, radionuclide bone imaging can be achieved, without exposing the patient to an unduly large dose of ionising radiation.

In some work, it is advantageous to use a β^+-emitting radionuclide, for the reason that when the positron undergoes mutual annihilation with an electron, the two 0.51 MeV γ-rays that ensue are emitted at the same instant of time and in mutually opposed directions. With sophisticated detection equipment, this allows the distribution of the radionuclide in the patient's body to be determined with greater accuracy than can be achieved with a γ-emitter.

One nuclide that may be utilised in this way, in positron emission tomography, is ^{18}F, with a half-life of 110 min. If these radioactive F atoms are attached to glucose molecules in a particular way, then partial metabolism of the glucose will occur, with the labelled molecule becoming trapped within the cell. Determining its location can then, for example, map metabolic activity within the brain. Another important application is in the development of tumours, since enhanced glucose metabolism is characteristic of tumour development and growth.

1.9 The chemical combination of atoms

The capacity of atoms to join together in simple ratios has been referred to earlier. When they so combine, we say that chemical bonds are formed between the atoms and the resulting entity is called a molecule. Following the suggestion of Couper in 1858, it is conventional to represent the bond by a line joining the two atoms. Thus the molecule HF, hydrogen fluoride, may be written as H–F: likewise water, H_2O, may be written H–O–H. The implication is that, on account of this chemical bond, the atoms are now attached to each other so that they move in concert, with some notional distance separating the nuclei.

If we consider also the hydrides of the two elements, N and C, coming before O and F in the second row of the Periodic Table, their formulae, NH_3 and CH_4, indicate that these four elements have varying capacities to combine with hydrogen. The term used here is valency (Lat: *valeo* = I am worth). We say that fluorine has a valency of one because an F atom can combine with only one hydrogen atom and it may participate in only one chemical bond. On the other hand, nitrogen has a valency of three because an N atom can combine with three hydrogen atoms and may participate in three chemical bonds. Similarly, the valencies of oxygen and carbon are respectively two and four.

One shortcoming of Dalton's atomic theory was that he did not envisage that atoms of the same element could be bonded together to form a molecule. Later, Avogadro appreciated that they do, so that the valency of the atom may be satisfied in the elemental state. Thus we have the molecules H_2 (H–H), F_2 (F–F), O_2 (O=O) and N_2 (N≡N) as the normal and stable form of these elements, where the symbol '=' denotes a double bond and '≡' a triple bond.

While each element has a normal valency, in certain instances it may exhibit a different valency. For example, the preferred and almost universal valency of carbon is four, as is illustrated by the molecules CH_4 and CF_4. It is also demonstrated by the reaction of carbon with oxygen to give CO_2, since oxygen is divalent and the molecule can be written as O=C=O. However, carbon and oxygen may also combine to give carbon monoxide, CO, and in this case the atoms of carbon and oxygen clearly do not each possess their preferred valencies.

Where there are more than two constituent atoms in a molecule, another aspect of valency relates to the geometry of these atoms. In the case of a triatomic molecule, the issue is simply whether the three atoms (that is to say, the nuclei of the three atoms) lie in a straight line. In some cases, such as CO_2, they are collinear and in others, such as H_2O, they are not, but both molecules have a definite shape, in terms both of bond lengths and the bond angle. For polyatomic molecules, the directional nature of valency is even more significant. Ammonia, NH_3, is not only non-linear but also non-planar.

For many simple compounds of two elements, the nature of the molecule can readily be deduced from the molecular formula. Thus SO_2 denotes an atom of sulfur with two atoms of oxygen bonded to it. However, for larger molecules, such information may not suffice to specify a particular compound. The molecular formula, C_5H_{12}, may describe any one of three different molecules, whose structures are:

(a)   (b)   (c)

In (a), three carbon atoms have two other C atoms bonded to them, but none has more than two. In (b), one carbon atom has three other C atoms bonded to it, while in (c), one carbon atom has the four other C atoms bonded to it. So structures (a), (b) and (c), all with the molecular formula, C_5H_{12}, are described as isomers. None of these molecules is planar, so that drawings such as the above on a plane surface are potentially misleading if too much detail is inferred from them.

For compounds containing more than two different elements, it is even easier to find examples of molecules with the same molecular formula but totally different structures. Consequently, if the name of the compound needs to be supported by a formula, something more specific than the molecular formula needs to be supplied. For example, one of the several compounds described by the formula $C_4H_8O_2$ is ethyl acetate, $CH_3CO_2C_2H_5$. This is a totally different substance from 1-butanoic acid, $CH_3(CH_2)_2CO_2H$, though they share this molecular formula.

Suggested reading

LOGAN, S. R., 1994, Instability of large nuclides with respect to decay by α-particle emission, *Journal of Chemical Education*, **71**, 888–89.

FRIEDLANDER, G., KENNEDY, J. W., MACIAS, E. S. and MILLER, J. M., 1981, *Nuclear and Radiochemistry*, 3rd Edn, New York: Wiley.

BERNIER, D. R., CHRISTIAN, P. E. and LANGAN, J. K. (Eds), 1994, *Nuclear Medicine: Technology and Techniques*, 3rd Edn, St Louis: Mosby.

McCARTHY, T. J., SCHWARZ, S. W. and WELCH, M. J., 1994, Nuclear medicine and positron emission tomography: an overview, *Journal of Chemical Education*, **71**, 830–36.

Problems

1.1 Given the masses, in a.m.u., of the following atoms:

^{1}H   =   1.0078252

^{60}Ni   =   59.93078

^{102}Ru   =   101.90434

and of the neutron

n   =   1.0086654

determine which nuclide, ^{60}Ni or ^{102}Ru, has:

(i) the greater nuclear binding energy, and
(ii) the greater nuclear binding energy per nucleon.

1.2 The nuclide ^{35}S decays by β^- emission:

$$^{35}_{16}S \rightarrow \; ^{35}_{17}Cl \; + \; ^{\;0}_{-1}\beta$$

Given the following atomic masses,

^{35}S   =   34.989034

^{35}Cl   =   34.968854

evaluate the maximum kinetic energy for the β^--particle emitted in this process.
(It is helpful to use the conversion factor, 1 a.m.u. $\equiv 931.4$ MeV)

1.3 The half-life of ^{35}S is 86.7 days. What fraction of a sample of this nuclide would survive for 1 year?

2

The extra-nuclear electrons

The arrangement of electrons within an atom is shown to be governed not by the mechanics of Newton, but by twentieth century quantum theory. Bohr introduced the idea that electrons would tend to occupy the states of lowest energy. Thus, in accordance with the Schrödinger equation, electrons occupy discrete energy levels, subject to the relevant rules.

Certain features in the electronic configuration of an atom are shown to recur with increasing atomic number. Since the chemical behaviour of an element is very much a function of the state of its highest energy electrons, this means that there is sufficient pattern to the behaviour of the elements to justify arranging them in the Periodic Table.

2.1 The Bohr atom

Shortly after Rutherford's discovery of the atomic nucleus, a model of the atom was proposed by Bohr, a Danish physicist. While this theory has long been superseded, it possesses several notable features which deserve to be mentioned.

Bohr conceived of the atom as being like a miniature solar system with the nucleus surrounded by orbiting electrons. Each electron was thought of as occupying a stable orbit, subject to certain criteria, in which the centrifugal force needed to maintain the electron in that orbit was exactly matched by the Coulombic force of attraction between the negatively charged electron and the positively charged nucleus. The orbit of lowest energy was therefore closest to the nucleus, with the energy progressively increasing with increasing radius of the orbit.

In the hydrogen atom, with only one electron, this would, in the ground state atom, occupy the lowest energy orbit available. In the case of other atoms it was necessary to invoke further arbitrary rules as to how many electrons could be accommodated in a particular orbit. Any atom may, of course, be elevated to an excited state in which the electrons possess greater energy than the minimum value corresponding to the ground state.

Two features of the Bohr atom deserve particular attention. Firstly, on the basis of the assumptions that were introduced, the radius of the orbit of the electron in the ground state hydrogen atom ('the Bohr radius') came to 0.529 Å or 0.529×10^{-10} m. Clearly, this value is extremely realistic. The second matter relates to the properties of light and it is appropriate first to reiterate two basic points, both known at that time. Light has wave properties, so that the velocity, c, is the product of the wavelength, λ, and the frequency, v, leading to the equation,

$$c = \lambda v \tag{2.1}$$

Also, light consists of discrete 'packets' called *quanta*, and the energy, ε, of a quantum is proportional to the frequency of the light:

$$\epsilon = hv \tag{2.2}$$

This equation was put forward by Planck in 1900 and the constant of proportionality, h, is called Planck's constant.

From the Bohr model it was possible to interpret well-known series of emission lines obtained by passing an electric current through hydrogen gas at a reduced pressure. Some of these lines may be attributed to emissions from individual hydrogen atoms, as the electron moves from a higher to a lower energy orbit.

The applicable equation is,

$$\epsilon_1 - \epsilon_2 = hv \tag{2.3}$$

where ϵ_1 and ε_2 denote the energies of the initial and final states.

The lines known as the Balmer Series lie in the visible range and the first four have wavelengths of 656.1, 486.0, 433.9 and 410.1 nm. These may be thought of as arising when electrons in yet higher energy orbits fall down to the first excited orbit, as sketched in Figure 2.1.

From the assumptions introduced by Bohr, the frequencies of these lines should satisfy the formula,

$$v = R_H \left(\frac{1}{2^2} - \frac{1}{N^2} \right) \tag{2.4}$$

where N has integral values higher than 2 and R_H represents a fundamental constant. To test this equation, the frequencies of the above lines, given by c/λ, are plotted against $1/N^2$, for $N = 3, 4, 5$ and 6. The linearity of the resulting plot serves to vindicate equation (2.4).

However, while a scientific theory that fails to meet a test may fairly be discarded, one which meets it does not necessarily deserve the gold seal of approval. Despite this apparent success of the Bohr atom, the model suffers from serious flaws.

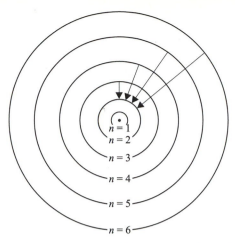

Figure 2.1 Diagrammatic representation of the first four lines of the Balmer Series, in terms of Bohr's model of the hydrogen atom.

2.2 The new quantum theory

In the years after Bohr made his proposal, a quiet revolution took place within physics in regard to the behaviour of small particles. The laws of Newtonian physics, which served so well for bodies of laboratory or of cosmic dimensions, were realised not to be applicable to subatomic particles, and especially to the electron.

The highlight of this revolution is contained within the Heisenberg Uncertainty Principle, which may be summarised by the statement that it is impossible, at the same time, to specify both the position and the momentum of a particle. The relevant equation is,

$$\Delta x . \Delta p = h/2\pi \tag{2.5}$$

where Δx and Δp denote the uncertainties in the position and the momentum, and h is Planck's constant.

Since the momentum is the product of the velocity and the mass, this limitation is not at all restrictive regarding the whereabouts of those track-and-field projectiles, the shot, the hammer and the discus. It is, however, extremely important in regard to the behaviour of an electron, whose mass is just under 10^{-30} kg. So while the behaviour of a shot-putter's projectile will quite adequately be described by Newton's laws of motion, that of an electron will not. As is illustrated in Figure 2.1, the Bohr atom assumes that an electron can be in a prescribed place travelling at a prescribed speed. In the light of the Uncertainty Principle, this assumption is unacceptable.

For the electron, the appropriate relation was found to be the Schrödinger equation. While we will omit any mention of the mechanics of solving this equation, we must look carefully at what the solutions indicate to us. In the

Schrödinger equation, the variable used is traditionally denoted by the Greek letter, ψ, and is called the **wave function**. (The inclusion of the word 'wave' arises from historical reasons, connected with the fact that the equation in question resembles those that describe certain types of wave motion.)

An important point is that there is not usually a unique solution for ψ, but rather a series of possible solutions, each compatible with the specified conditions. Thus the electron may occupy one of several discrete energy levels, which means that the Schrödinger equation reproduces this major assumption of the Bohr theory. The wave function, ψ, may be either positive or negative. The physical significance of ψ is that the value of ψ^2 at any point is a measure of the probability of finding the electron there. (Note that ψ^2 is always positive!)

The picture which emerges is of the electron having some vestige of a presence throughout a finite volume. Over a period of time, the probability of finding the electron at a certain point may be equated to the electron density at that point, so that the electron is not a localised particle, but is in effect smeared over a significant volume. The name that is used to denote this spatial disposition of the electron is 'orbital', which reflects the enduring contribution of the Bohr atom.

2.3 The Schrödinger equation and the hydrogen atom

The Schrödinger equation appropriate to the hydrogen atom is technically simple, since there is only one nucleus and one electron which will exert on each other the usual forces of Coulombic attraction. Exact solutions are therefore possible.

The orbital of lowest energy is a spherically symmetrical one, in which ψ decreases with increasing separation, r, from the nucleus, as is shown in Figure 2.2(a). Thus the electron density is highest at the nucleus. However, there is only one point at the position $r = 0$, but innumerable points at a separation $r = r$, so that there must, in total, be more electron density at finite values of r than there is at $r = 0$. Since the surface area of a sphere of radius r is $4\pi r^2$, and the electron density is proportional to the square of the wave function, the relevant function is $4\pi r^2 \psi^2$. The plot of $4\pi r^2 \psi^2$, shown in Figure 2.2(b), demonstrates that the total electron density function rises to a maximum and then declines. This maximum occurs at 0.529 Å, a distance identical to the Bohr radius (see section 2.1).

One feature of this orbital is that the wave function ψ is everywhere positive, tending towards zero as $r \to \infty$. This means that, uniquely for the 1s orbital, the electron density is finite at every point. For all the other orbitals, ψ undergoes one or more changes in sign and in the process passes through zero. Such a location is described as a **node**, a term employed in regard to standing waves, such as the vibrations of a violin string, for a site of perpetually zero displacement.

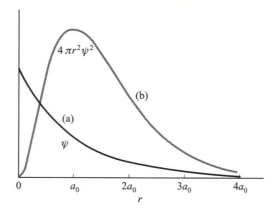

Figure 2.2   Plots of ψ and of $4\pi r^2\psi^2$ as functions of r for the 1s orbital of the hydrogen atom.

The labels given to identify individual orbitals are made up of a combination of a simple integer denoting the principal quantum number, n, and (for obscure historical reasons) one of the lower case letters, s, p, d and f. On that basis, the orbital described above and depicted in Figure 2.2 is called the 1s orbital. All s orbitals are spherically symmetrical, but whereas the 1s orbital has no nodes, the 2s has one spherical node and the 3s has two spherical nodes. In all cases the value of ψ and thus of the electron density is finite at the nucleus.

For the principal quantum number n equal to 1, the only possible orbital is the 1s. When $n = 2$, then in addition to the 2s orbitals there are the 2p orbitals, which have a nodal plane passing through the nucleus and are symmetrical along an axis at right angles to this plane. Since this axis may be in any one of three mutually perpendicular directions, there are in fact three such orbitals, called the $2p_x$, the $2p_y$ and the $2p_z$ orbitals, all of the same energy.

When $n = 3$, then in addition to the 3s and the three mutually perpendicular 3p orbitals, there also exist orbitals of another type, designated 3d. The number of distinct 3d orbitals, all of equivalent energy, is five. Each of these orbitals of principal quantum number 3 has two nodal surfaces, illustrating the general principle that the energy levels of orbitals rise with increasing numbers of nodes.

2.4   The shapes of atomic orbitals

The difficulties of representing, on a plane surface, the distribution of electron density in an atomic orbital have been touched on in the previous section, particularly in regard to orbitals of spherical symmetry. For orbitals that do not have spherical symmetry the problem is even more awkward since more than one co-ordinate is now required to represent position.

For the 2s orbital, the variation of the wave function and of $4\pi r^2\psi^2$ with r is shown in Figure 2.3. The highest value of $4\pi r^2\psi^2$ lies outside the spherical node. Likewise, for the 3s orbital, shown in Figure 2.4, the highest value of $4\pi r^2\psi^2$ lies outside the outer spherical node. Thus the preponderance of the electron density is at some distance from the nucleus for both these orbitals.

It is sufficient to represent one of the 2p orbitals. The variations of the wave function of the $2p_x$ orbital along the x axis, and of ψ^2, are shown in Figure 2.5. There is a node in the yz plane, so that the function ψ^2 is symmetrical about this nodal plane, with the maximum electron density being achieved at distances of $1.06\,\text{Å}$ from the nucleus. Another way to depict this orbital, to show

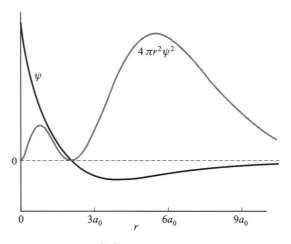

Figure 2.3   Plots of ψ and of $4\pi r^2\psi^2$ as functions of r for the 2s orbital of the hydrogen atom.

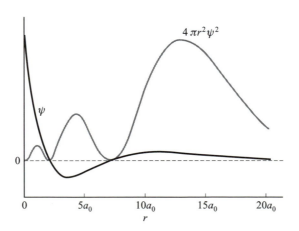

Figure 2.4   Plots of ψ and of $4\pi r^2\psi^2$ for the 3s orbital of the hydrogen atom.

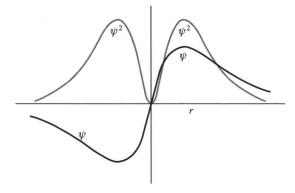

Figure 2.5   Plots of ψ and of ψ^2 along the x axis for the 2p$_x$ orbital of the hydrogen atom.

its disposition around the x axis, is to draw contours of electron density in a plane through this axis. This is shown in Figure 2.6. The signs $+$ and $-$ are added to emphasise the fact that between the two lobes of electron density there lies a nodal plane, across which the wave function changes sign.

The subdivision of the five 3d orbitals is more complex. These are given the subscripted labels 3d$_{xy}$, 3d$_{yz}$, 3d$_{xz}$, 3d$_{x^2-y^2}$ and 3d$_{z^2}$. The first three all lie in and are symmetrical about the designated plane. Each of these orbitals has two mutually perpendicular nodal planes passing through the nucleus, as is shown in Figure 2.7. The 3d$_{x^2-y^2}$ orbital is broadly similar in shape to these three,

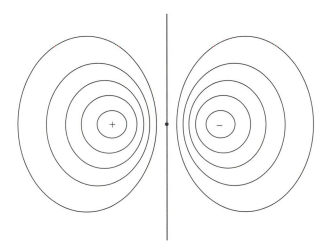

Figure 2.6   Contours of electron density in the xy plane for the 2p$_x$ orbital of the hydrogen atom.

while the $3d_{z^2}$ orbital is totally different, being symmetrical about the z axis, with two cone-shaped nodal surfaces. These also are depicted in Figure 2.7.

2.5 Electronic states in polyelectronic atoms

In atoms other than the hydrogen atom, the solutions to the Schrödinger equation referred to in the previous section are not strictly correct in that these have been derived by neglecting the repulsive interactions between pairs of electrons. Exact solutions reflecting these repulsions are not feasible, but it is not difficult to achieve acceptable approximations.

As is illustrated in Figures 2.2, 2.3, 2.4 and 2.5, most of the electron density of the 1s orbital lies much closer to the nucleus than does that of the 2s or the 2p orbitals. This means that an electron occupying a 1s orbital tends to screen any other electron from the nuclear charge and so its main effect is to alter the dimensions but not the general shape of the orbitals of principal quantum number 2 or 3. Likewise, most of the electron density of orbitals with $n = 2$ lies well inside that of orbitals with $n = 3$. Thus, for atoms containing several electrons, the relevant orbitals are of the same types as those detailed in the previous section in regard to the hydrogen atom.

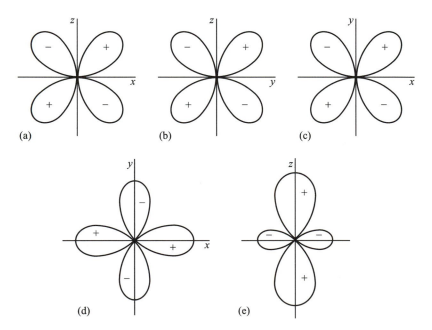

Figure 2.7   Sketches of the disposition of the electron density of the five 3d orbitals of the hydrogen atom: (a) d_{xz} (b) d_{yz} (c) d_{xy} (d) $d_{x^2-y^2}$ and (e) d_{z^2}.

An additional factor of some importance is the existence of electron spin, first proposed to explain small splittings observed in the spectral lines of certain atoms. The initial discussion of electron spin reflected the fact that the electron was then regarded as a particle, but accepting its true quantum mechanical nature, an electron has a spin quantum number, m_s, of either $+\frac{1}{2}$ or $-\frac{1}{2}$ in the appropriate units.

The electronic configurations of isolated atoms may be explained in terms of three principles. The first of these, usually called the **Aufbau principle** (from the German word for 'construction') was actually introduced by Bohr. It says, in effect, that the electronic make-up of a ground state atom may be replicated if one allocates electrons, in turn, each to the lowest energy orbital available. In applying it, one needs also to use the **Pauli exclusion principle**, which is that no two electrons may have identical quantum numbers. Thus if the spin quantum number of the first electron in the 1s orbital is $+\frac{1}{2}$, that of the second must be $-\frac{1}{2}$. At that point, the 1s orbital is fully subscribed and, applying the Aufbau principle, the next electron must go into the 2s orbital.

On that basis we may write down the electronic configurations of isolated ground state atoms of the first four elements:

$$Z = 1 \quad \text{H} \quad 1s^1$$
$$2 \quad \text{He} \quad 1s^2$$
$$3 \quad \text{Li} \quad 1s^2\,2s^1$$
$$4 \quad \text{Be} \quad 1s^2\,2s^2$$

Here, the superscript after the designation of the atomic orbital indicates the number of electrons occupying that orbital. Where the superscript is 2, this means that for one electron, $m_s = +\frac{1}{2}$ and for the other, $m_s = -\frac{1}{2}$: that is, the electron spins are paired.

In the hydrogen atom, the energies of the 2s and the 2p orbitals are equal. However, in the presence of other electrons, this balance is disturbed. Where the 1s orbitals are occupied, the energy level of the 2s orbitals lies a little way below that of the three 2p orbitals which, as explained on p. 23, all have the same energy and are consequently termed **degenerate** orbitals.

The third principle, called **Hund's rule of maximum multiplicity**, has the implication that when the number of electrons occupying the p orbitals lies between two and four, then the electrons are disposed so that as many as possible are unpaired, with parallel spins. So the ground states of the next six elements are:

$$Z = 5 \quad \text{B} \quad 1s^2 \quad 2s^2 \quad 2p_x^{\,1}$$
$$6 \quad \text{C} \quad 1s^2 \quad 2s^2 \quad 2p_x^{\,1} \quad 2p_y^{\,1}$$
$$7 \quad \text{N} \quad 1s^2 \quad 2s^2 \quad 2p_x^{\,1} \quad 2p_y^{\,1} \quad 2p_z^{\,1}$$
$$8 \quad \text{O} \quad 1s^2 \quad 2s^2 \quad 2p_x^{\,2} \quad 2p_y^{\,1} \quad 2p_z^{\,1}$$
$$9 \quad \text{F} \quad 1s^2 \quad 2s^2 \quad 2p_x^{\,2} \quad 2p_y^{\,2} \quad 2p_z^{\,1}$$
$$10 \quad \text{Ne} \quad 1s^2 \quad 2s^2 \quad 2p_x^{\,2} \quad 2p_y^{\,2} \quad 2p_z^{\,2}$$

(It must be admitted that the practice of assigning electrons to the $2p_x$ before the $2p_y$ before the $2p_z$ orbital is totally arbitrary. In that regard, these assignments for B, C, O and F cannot be defended to the death.)

2.6 The emergence of the Periodic Table

During the nineteenth century, as increasing numbers of chemical elements were discovered and their atomic masses determined, attempts were made to see whether these elements belonged to any discernable pattern. To this end, Newlands arranged the known elements in the order of increasing atomic mass and found some recurring pattern, summarised in his Law of Octaves. His critics said cynically that he might as well have put them in alphabetical order.

These early attempts foundered largely because of two difficulties which were then unsuspected. One is that certain elements were then unknown. Also, from a chemical point of view, there is nothing really fundamental about the atomic mass. The crucially important parameter belonging to an element is its atomic number, Z. These two tend to increase together, as can be seen from Table 1.1. However, there are instances where an element with a higher atomic number has a lower value of the atomic mass. Consequently, ordering the elements in terms of their increasing atomic mass does not entirely achieve a listing in terms of increasing atomic number.

The first satisfactory arrangement of the elements in terms of their chemical behaviour is usually credited to Mendeleev, a Russian chemist, in 1869. At that time, the concept of atomic number was unknown, but Mendeleev was confident that, in certain instances, using increasing atomic mass resulted in the elements being in the wrong order. So he put tellurium before iodine, and cobalt before nickel.

His intuition also led Mendeleev to assert that there were unknown chemical elements which might be discovered in the future, so he confidently left spaces for these. As a result, he achieved a listing of elements showing a recurring periodicity in their behaviour and this has become known as the Periodic Table. The subsequent discovery of the elements scandium, gallium, germanium and polonium, with properties very close to those predicted by Mendeleev, added to the lustre of his achievement.

Whereas in 1869, only about 65 elements were known, the current number is 115. So the Periodic Table has evolved, not least because of the discovery of the various noble gases. There has been (and still is) continual debate on how one should write the Periodic Table. The version shown in Figure 2.8 incorporates features which are widely accepted and serves to demonstrate the connection between electronic configuration and chemical behaviour.

The Periodic Table of the Elements (Figure 2.8). Group headings run I, II, (Transition elements), III, IV, V, VI, VII, O across Rows 1–7.

Group I	II				Transition elements							III	IV	V	VI	VII	O
Row 1 — 1 H $1s^1$																	2 He $1s^2$
Row 2 — 3 Li $(He)2s^1$	4 Be $(He)2s^2$											5 B $(He)2s^22p^1$	6 C $(He)2s^22p^2$	7 N $(He)2s^22p^3$	8 O $(He)2s^22p^4$	9 F $(He)2s^22p^5$	10 Ne $(He)2s^22p^6$
Row 3 — 11 Na $(Ne)3s^1$	12 Mg $(Ne)3s^2$											13 Al $(Ne)3s^23p^1$	14 Si $(Ne)3s^23p^2$	15 P $(Ne)3s^23p^3$	16 S $(Ne)3s^23p^4$	17 Cl $(Ne)3s^23p^5$	18 Ar $(Ne)3s^23p^6$
Row 4 — 19 K $(Ar)4s^1$	20 Ca $(Ar)4s^2$	21 Sc $(Ar)4s^23d^1$	22 Ti $(Ar)4s^23d^2$	23 V $(Ar)4s^23d^3$	24 Cr $(Ar)4s^13d^5$	25 Mn $(Ar)4s^23d^5$	26 Fe $(Ar)4s^23d^6$	27 Co $(Ar)4s^23d^7$	28 Ni $(Ar)4s^23d^8$	29 Cu $(Ar)4s^13d^{10}$	30 Zn $(Ar)4s^23d^{10}$	31 Ga $(Ar)4s^23d^{10}4p^1$	32 Ge $(Ar)4s^23d^{10}4p^2$	33 As $(Ar)4s^23d^{10}4p^3$	34 Se $(Ar)4s^23d^{10}4p^4$	35 Br $(Ar)4s^23d^{10}4p^5$	36 Kr $(Ar)4s^23d^{10}4p^6$
Row 5 — 37 Rb $(Kr)5s^1$	38 Sr $(Kr)5s^2$	39 Y $(Kr)5s^14d^1$	40 Zr $(Kr)5s^24d^2$	41 Nb $(Kr)5s^24d^4$	42 Mo $(Kr)5s^14d^5$	43 Tc $(Kr)5s^14d^6$	44 Ru $(Kr)5s^14d^7$	45 Rh $(Kr)5s^14d^8$	46 Pd $(Kr)5s^04d^{10}$	47 Ag $(Kr)5s^14d^{10}$	48 Cd $(Kr)5s^24d^{10}$	49 In $(Kr)5s^24d^{10}5p^1$	50 Sn $(Kr)5s^24d^{10}5p^2$	51 Sb $(Kr)5s^24d^{10}5p^3$	52 Te $(Kr)5s^24d^{10}5p^4$	53 I $(Kr)5s^24d^{10}5p^5$	54 Xe $(Kr)5s^24d^{10}5p^6$
Row 6 — 55 Cs $(Xe)6s^1$	56 Ba $(Xe)6s^2$	57 La $(Xe)6s^25d^1$	72 Hf $(Xe)6s^24f^{14}5d^2$	73 Ta $(Xe)6s^24f^{14}5d^3$	74 W $(Xe)6s^24f^{14}5d^4$	75 Re $(Xe)6s^24f^{14}5d^5$	76 Os $(Xe)6s^24f^{14}5d^6$	77 Ir $(Xe)6s^24f^{14}5d^7$	78 Pt $(Xe)6s^14f^{14}5d^9$	79 Au $(Xe)6s^14f^{14}5d^{10}$	80 Hg $(Xe)6s^24f^{14}5d^{10}$	81 Tl $(Hg)6p^1$	82 Pb $(Hg)6p^2$	83 Bi $(Hg)6p^3$	84 Po $(Hg)6p^4$	85 At $(Hg)6p^5$	86 Rn $(Hg)6p^6$
Row 7 — 87 Fr $(Rn)7s^1$	88 Ra $(Rn)7s^2$	89 Ac $(Rn)7s^26d^1$	104 Rf	105 Db	106 Sg	107 Bh	108 Hs	109 Mt									

Lanthanide series

58 Ce $(Xe)6s^24f^2$	59 Pr $(Xe)6s^24f^3$	60 Nd $(Xe)6s^24f^4$	61 Pm $(Xe)6s^24f^5$	62 Sm $(Xe)6s^24f^6$	63 Eu $(Xe)6s^24f^7$	64 Gd $(Xe)6s^24f^75d^1$	65 Tb $(Xe)6s^24f^9$	66 Dy $(Xe)6s^24f^{10}$	67 Ho $(Xe)6s^24f^{11}$	68 Er $(Xe)6s^24f^{12}$	69 Tm $(Xe)6s^24f^{13}$	70 Yb $(Xe)6s^24f^{14}$	71 Lu $(Xe)6s^24f^{14}5d^1$

Actinide series

90 Th $(Rn)7s^26d^2$	91 Pa $(Rn)7s^26d^15f^2$	92 U $(Rn)7s^26d^15f^3$	93 Np $(Rn)7s^26d^15f^4$	94 Pu $(Rn)7s^26d^15f^5$	95 Am $(Rn)7s^25f^7$	96 Cm $(Rn)7s^26d^15f^7$	97 Bk $(Rn)7s^25f^86d^1$	98 Cf $(Rn)7s^25f^{10}$	99 Es $(Rn)7s^25f^{11}$	100 Fm $(Rn)7s^25f^{12}$	101 Md $(Rn)7s^25f^{13}$	102 No $(Rn)7s^25f^{13}6d^1$	103 Lr

Figure 2.8 The Periodic Table of the Elements.

2.7 The Periodic Table and electronic configuration

Hydrogen, as the first element, is understandably anomalous. It does not chemically resemble any other element. Helium, with a full complement of electrons with the quantum number $n = 1$, with the configuration $1s^2$, is chemically inert and belongs to Group O. Thereafter, while new electrons are being added, successively, to the 2s, 2p, 3s and 3p orbitals, the elements show periodicity on the basis of the number of electrons with the greatest principal quantum number. Thus lithium, with the configuration $1s^2 2s^1$ which we may more briefly write as $(He)2s^1$, is like sodium, $(Ne)3s^1$. Likewise fluorine, $(He)2s^2 2p^5$ is like chlorine, $(Ne)3s^2 3p^5$. Neon, with $1s^2 2s^2 2p^6$, has all the electrons possible for $n = 2$ and resembles helium, as does argon, with $1s^2 2s^2 2p^6 3s^2 3p^6$.

In the excited hydrogen atom, with no inter-electron interactions, the energy levels of the 3s, 3p and 3d orbitals are all equal. However, when the orbitals of lower energy than these are all occupied, not only do the energies of these orbitals increase in that order, but the energy of the 3d orbitals lies above that of the 4s orbitals. So the next element after argon, namely potassium, has the electronic structure $(Ar)4s^1$, and chemically resembles sodium. The one after, calcium, has the structure $(Ar)4s^2$ and resembles magnesium.

By the Aufbau principle, the next electrons are allocated to the 3d orbitals. So in the next 10 elements, the electrons of highest energy are in a 3d orbital, but the outermost electrons occupy the 4s orbital. In consequence, their chemical properties are rather different from those of any elements encountered previously. They are collectively known as transition elements and, as shown in Figure 2.8, they are usually allocated a distinctive place in the Periodic Table.

After zinc, the next electrons are allocated to the 4p orbitals, so the six elements from gallium to krypton resemble those from aluminium to argon. The electronic configurations of the next 18 elements after krypton are entirely analogous to those from potassium to krypton, with the sole difference that every principal quantum number has been increased by one.

For the principal quantum number equal to 4, then in addition to the 4s, 4p and 4d orbitals, there are also seven degenerate 4f orbitals. When the orbitals of lower energy are fully occupied, then the energies of the above orbitals increase in the order in which they have been listed. In fact, the 6s orbitals lie lower in energy than do the 4f orbitals, which are very close in energy to the 5d orbitals. So after xenon ($Z = 54$), the next two elements, caesium and barium, have the configurations $(Xe)6s^1$ and $(Xe)6s^2$, and resemble potassium and calcium.

The next element, lanthanum, has the configuration $(Xe)6s^2 5d^1$, but thereafter the electrons additional to those of barium occupy the 4f orbitals. So the elements from $Z = 58$ to 71 inclusive do not belong to any of the Groups specified to date and are known as the Lanthanide Series. From $Z = 72$ onwards, with the 4f orbitals full, additional electrons are allocated to the 5d orbitals and these elements constitute the third series of the transition elements.

After the 5d orbitals, the 6p orbitals are predictably filled next. By this stage, at $Z = 86$, we have exceeded the value of the maximum atomic number for which any stable nuclides exist. However, the half-lives of many of the naturally occurring unstable nuclides are quite long so that, for these elements at least, radioactivity is not a barrier to a study of the chemistry. Basically, the use of the 7s, the 5f and the 6d orbitals is largely analogous to the process described above for the orbitals with principal quantum numbers one unit less. After $Z = 87$, we have two elements resembling caesium and barium, one resembling lanthanum, followed by the Actinide Series which is analogous to the Lanthanide Series, and then a fourth series of transition elements. By that stage, we have attained an atomic number of 112.

The latter part of this account has been of the broad-brush variety and has not made mention of any of the subtleties of electronic configuration found among the transition elements. For example, whereas titanium and vanadium have the configurations $(Ar)4s^2\ 3d^2$ and $(Ar)4s^2\ 3d^3$, the next element, chromium, has $(Ar)4s^1\ 3d^5$. Five places further on, copper has $(Ar)4s^1\ 3d^{10}$.

In filling the 3d orbitals, which are degenerate in an isolated atom, Hund's rule applies, just as it does to the 2p and the 3p orbitals. So at the beginning of the transition element series, one electron goes into each of these five 3d orbitals. The anomaly at chromium may be put down to the particular stability of having all five 3d orbitals singly occupied. Manganese then has $(Ar)4s^2\ 3d^5$. Similarly there is special stability in having all five orbitals doubly occupied, which explains the behaviour of copper.

While the comprehensive details of the various apparent anomalies in electronic configuration are not specially relevant in an introductory account, it is desirable to identify their origins. With rising atomic number, increasingly there are involved various atomic orbitals whose energy levels lie close together. While the Aufbau principle is fully applicable, irrespective of how close in energy these levels lie, it has to be remembered that the precise energy levels of the various atomic orbitals are affected by the state of occupancy of these and other orbitals.

2.8 Ionisation behaviour of the individual atom

To illustrate the properties of one atom of an element, it is useful to look at the respective energies involved in carrying out the operations which (i) remove one electron from an isolated ground state atom and (ii) add an additional electron to such an atom. These quantities show periodicities which are totally consistent with the arrangement of the elements in the Periodic Table.

For the hypothetical element E, the **ionisation potential** (IP) denotes the minimum energy required to achieve the process:

$$E(g) \quad \rightarrow \quad E^+(g) + e^-(g) \tag{2.6}$$

The variation of the ionisation potential with the atomic number, Z, is shown in Figure 2.9. Two features stand out from this diagram. It may be seen that the first element introducing a new principal quantum number (so that its electronic configuration ends in ns^1) has a specially low value of the ionisation potential. These elements, Li, Na, K, Rb and Cs, constitute Group I and are known as the alkali metals. The low IP value is to be correlated with the fact that the ns orbital extends far out from the nucleus so that an electron which occupies this orbital is easily removed.

Secondly, the element immediately before this has a notably high value of the ionisation potential. These elements, He, Ne, Ar, Kr and Xe make up Group O, called the noble gases. In the case of helium the electrons of highest energy are in a 1s orbital, whereas in every other case the noble gas is reached when the p orbitals have been filled, regardless of whether there are d orbitals of the same principal quantum number. Also, as the principal quantum number increases, the value of the maximum ionisation potential achieved at Group O steadily declines.

The other parameter regarding an element is the **electron affinity** (EA), defined as the minimum energy required to carry out the process:

$$E^-(g) \quad \rightarrow \quad E(g) + e^-(g) \tag{2.7}$$

This means that the electron affinity is a measure of the tendency for an atom to acquire an additional electron. The dependence of the electron affinity on the atomic number is depicted in Figure 2.10. The salient feature of this plot is the consistently high value for the element immediately before a noble gas, with the exception of H which is anomalous in many ways. These elements invariably have five electrons in their p orbitals, constitute Group VII and are called the halogens. While the EA value decreases from chlorine to bromine to iodine,

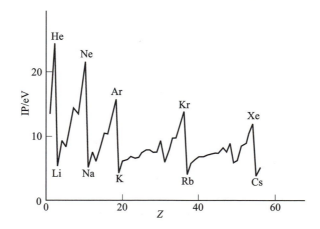

Figure 2.9 A plot of ionisation potential against atomic number for $1 \leq Z \leq 56$.

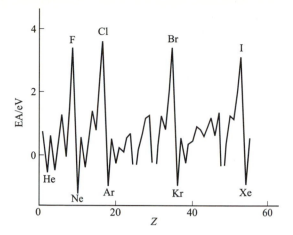

Figure 2.10 A plot of electron affinity against atomic number for $1 \leq Z \leq 56$.

that of fluorine does not fit the general pattern so that the element with the highest electron affinity is chlorine. The elements in Group VI, immediately before the halogens, have fairly high values of electron affinity, but for several other elements, including the noble gases, where there is absolutely no tendency to acquire an extra electron, the value is negative.

Suggested reading

MURRELL, J. N., KETTLE, S. F. A. and TEDDER, J. M., 1985, *The Chemical Bond*, 2nd Edn, Chichester: Wiley.

McWEENY, R., 1979, *Coulson's Valence*, 3rd Edn, London: Oxford University Press.

KARPLUS, M. and PORTER, R. N., 1970, *Atoms and Molecules: An Introduction for Students of Physical Chemistry*, New York: Benjamin.

PUDDEPHATT, R. J. and MONAGHAN, P. K., 1986, *The Periodic Table of the Elements*, London: Oxford University Press.

WEBSTER, B. C., 1990, *Chemical Bonding Theory*, Oxford: Blackwell.

Problems

2.1 The Paschen Series, which is analogous to the Balmer Series mentioned in section 2.1, involves transitions from various N values to $N = 3$. Give the formula which, in place of Equation (2.3), will be applicable to the Paschen Series. Also, calculate the wavelengths of the first three lines of the Paschen Series.

2.2 For a hydrogenic atom (i.e. one with only one electron) of atomic number Z, the most probable radius of the location of the electron in a 1s orbital is given by:

$$r^* = a_0/Z$$

where a_0 denotes the Bohr radius of the hydrogen atom, equal to 0.529 Å. What then is the most probable radius of the ground state orbital of a single electron around the nucleus, $^{238}_{92}U$?

2.3 For the 1s orbital of the hydrogen atom, the wave function ψ is given by:

$$\psi = \left(\frac{1}{\pi a_0^3}\right)^{1/2} e^{-r/a_0}$$

where a_0 denotes the Bohr radius. Evaluate the relative values of ψ and of ψ^2 at $r = 0$, $r = a_0$, $r = 2a_0$, $r = 3a_0$ and $r = 4a_0$. Also calculate the relative values of $4\pi r^2 \psi^2$ at these distances.

3

Chemical bonding

The ability of atoms to combine with one another to form molecules or for atoms of different elements to react to form a compound is directly attributable to the behaviour of their extra-nuclear electrons. While a pair of dissimilar elements may form an ionic compound by transferring an electron from one atom to the other to yield M^+X^-, the building up of discrete molecules, whether of a compound or of an element, is mostly achieved by electron sharing in what are called covalent bonds. These are described here in terms of the molecular orbital (MO) approximation. This may be seen as an extension of the idea of an atomic orbital disposed around one nucleus: the alternatives are a localised MO confined around two nuclei, or a delocalised MO around some greater number. While this treatment draws on the description of atomic orbitals in the previous chapter, it lays the basis for interpreting the ultraviolet spectra of molecules, discussed in the next chapter, in terms of transitions between different electronic states.

3.1 The ionic 'bond'

From quite early times, it was recognised that in some cases the formation of a chemical compound results from the interaction of elements of diametrically opposed tendencies. A good illustration of this phenomenon is provided by the alkali metal halides, formed between elements of Group I and Group VII of the Periodic Table.

As is shown in Figure 2.9, the alkali metals all have low ionisation potentials, and each member of this Group represents a minimum in the plot of IP against atomic number. Likewise, the unusually high electron affinities of the halogens are demonstrated in Figure 2.10. If an electron were to be transferred from an atom of an alkali metal, M, to a halogen atom, X, to create M^+ and X^-, these ions would exert mutual Coulombic attraction. The consequence for a large number of pairs of such ions would be that they achieve a state of lower energy by forming a lattice. In this way it is energetically possible for the ionic compound MX to be formed from the elements.

Not all the alkali metal halides adopt the same crystal structure. The structure adopted by sodium chloride, NaCl, is illustrated in Figure 3.1. In this, each Na^+ ion has, as its nearest neighbours, six Cl^- ions and each Cl^- ion has, as its nearest neighbours six Na^+ ions. That is, in the NaCl structure, there is a co-ordination number of six. Caesium chloride, CsCl, has a different crystal structure with a co-ordination number of eight. This latter structure, also illustrated in Figure 3.1, is to be preferred on energetic grounds, but NaCl is unable to conform to it simply because the Na^+ ion is so small that it is not possible for eight Cl^- ions to be fitted around it.

A significant consequence of forming a chemical compound in this way is that no discrete molecule of MX actually exists. In the solid state, an ion M^+ is related in just the same way to each of six (or eight) X^- ions. In the melt or in solution, each M^+ ion and each X^- ion moves around independently. To that extent, the term 'ionic bond' is a misnomer.

Ionic compounds may also be formed which involve elements of Group II, the alkaline earths, or of Group VI. While the ionisation potentials of the alkaline earths are not as low as those of the alkali metals, they are fairly low. In addition, the second ionisation potentials of these elements are quite low. Thus a magnesium atom might donate an electron to each of two chlorine atoms, to form a compound whose stoichiometric formula is $MgCl_2$.

Also, the electron affinity of sulfur, while not as high as that of chlorine, may be seen from Figure 2.10 to be quite high. It is also possible for the sulfur

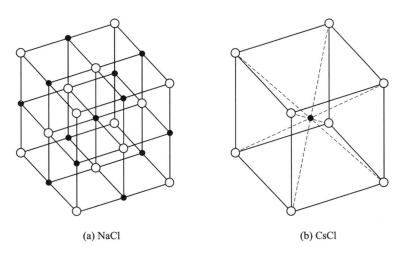

(a) NaCl   (b) CsCl

Figure 3.1   Illustration of the crystal structures of (a) sodium chloride, NaCl and (b) caesium chloride, CsCl.

atom to acquire a second electron, so that it may form a compound with an alkali metal of formula M_2S. Numerous other ionic oxides and sulfides are formed with elements of Group II or Group III.

3.2 The covalent bond: the molecular orbital approach

While the earliest conception of chemical combination was of the interaction of opposites, it soon became clear that similar and even identical atoms could combine to form a molecule. This behaviour was not anticipated by Dalton, who did not realise that hydrogen, for example, existed as diatomic H_2 molecules. In molecules of this type, the chemical bonds involve the sharing of electrons and are classed as covalent bonds.

To explain the covalent bond in terms of quantum theory, it is helpful to start with the hydrogen molecule ion, H_2^+. Mass spectrometric studies of the ionic species produced in hydrogen by electron impact or by a radiofrequency discharge show that this ion is stable in the gas phase. Moreover, although it is composed of three charged particles, two of these are nuclei which are relatively slow-moving and only one is an electron, so that for H_2^+, an exact solution of the Schrödinger equation is possible. This was achieved by Bates and co-workers in 1953.

There are two alternative solutions for the H_2^+ ion, as is illustrated in Figure 3.2. The orbital of lower energy is symmetrical about the plane midway between the two nuclei. The electron density is highest at the nuclei, but remains high in the region between the two nuclei. The presence of a significant

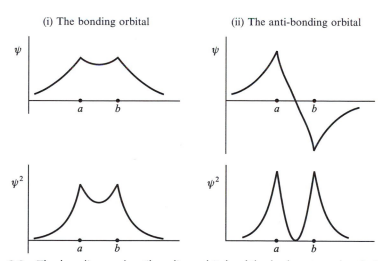

Figure 3.2 The bonding and antibonding orbitals of the hydrogen molecule ion, H_2^+. Plots of ψ and ψ^2 along the internuclear axis for (i) the bonding orbital and (ii) the antibonding orbital.

amount of electronic charge in this region tends to nullify the mutual repulsion of the two nuclei and thus assists in holding the ion together.

The other solution to the Schrödinger equation for H_2^+ is an orbital with a node in the plane midway between the two nuclei, also shown in Figure 3.2. Because of the node, this orbital has a higher energy than the first orbital, which has no node. The presence of a nodal plane means that the wave function, ψ, and the electron density fall to zero at the plane midway between the two nuclei, so this orbital is not conducive to holding the ion together.

The two orbitals illustrated in Figure 3.2 may be regarded as originating from the combination of the 1s orbitals around the two hydrogen nuclei. It is convenient to label these a and b. In the wave function for H_2^+, the coefficients for ψ_a and ψ_b should on symmetrical grounds be equal. However, in the aggregate orbital we may have either $(\psi_a + \psi_b)$ or $(\psi_a - \psi_b)$. The former corresponds to the orbital of lower energy and is called the bonding orbital. The latter, with a higher energy, is called the antibonding orbital, for the reason detailed above.

These new orbitals are normally called molecular orbitals (MO). The positions of their energies, relative to those of the component 1s atomic orbitals, are illustrated in Figure 3.3. This demonstrates that in the bonding MO, the greater delocalisation of the electron density than there is in either 1s orbital leads to a lowering in energy. The antibonding MO, with comparable delocalisation of the electronic charge, has a higher energy than the 1s orbitals because a node has now been introduced.

The degree of symmetry possessed by a molecular orbital is necessarily different from that of an atomic orbital. Whereas an atomic orbital may have spherical symmetry, a molecular orbital, around two (or more) nuclei, may at most have axial symmetry. So, to denote a molecular orbital with axial symmetry, we use the Greek letter *sigma* (σ), which corresponds to the Arabic

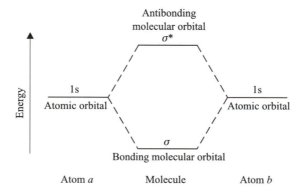

Figure 3.3 Diagram of the energies of the bonding and the antibonding orbitals of H_2^+ (and H_2) in relation to those of the 1s orbitals of the H atom.

letter s used to denote an atomic orbital of spherical symmetry. Of the two MOs shown in Figures 3.2 and 3.3, the bonding orbital is classified as σ and the antibonding one as σ^*.

For a molecular system with more than one electron, an exact solution of the Schrödinger equation is not possible. This means that, to describe the bonding in these systems, it is necessary to make approximations. The MO approach is to treat the molecular orbitals described above in a manner analogous to the use, in the previous chapter, of the atomic orbitals of hydrogen-like atoms for multi-electron atoms. Thus in the molecule H_2, one would expect that in the ground state, two electrons, with spins paired, would occupy the σ orbital in accordance with the Aufbau and the Pauli principles. The (hypothetical) molecule, He_2, has four electrons and these would be allocated, on the same basis, two to the σ and two to the σ^* orbital. As is shown in Figure 3.3, the latter orbital is antibonding to the same extent as the σ orbital is bonding. This means that there is no net binding force in He_2 and so this molecule does not exist.

3.3   The sigma bond in diatomic molecules

The alkali metals are solids at room temperature but they are fairly volatile and exist in the vapour phase as diatomic molecules, M_2. Let us look, in MO terms, at the electronic structure of Li_2.

Lithium, with $Z = 3$, has the electronic configuration, $1s^2\,2s^1$. If two such atoms interact, it might seem possible to form molecular orbitals both from the 1s and the 2s orbitals. However, a comparison of Figures 2.2 and 2.3 indicates that, if the nuclei are at a separation which is optimum for the formation of the σ_{2s} orbital, then there will be negligible interaction between the 1s orbitals of the two atoms. This means that in covalently bonded molecules of second row elements, the 1s electrons may be treated as 'core' electrons. So for bonding purposes we need consider only the 2s orbitals.

In the molecule Li_2, two electrons, with spins paired, occupy the σ_{2s} molecular orbital, formed by combination of the separate 2s orbitals of the two Li atoms. The molecule is thus closely analogous to H_2. In the molecule Na_2, the electrons of principal quantum numbers 1 and 2 are all treated as 'core' and the bonding MO is the σ_{3s} orbital, that is, the sigma orbital formed from the 3s atomic orbitals. Not surprisingly, the internuclear distance in Na_2, at 3.08 Å, is substantially greater than that in H_2 (0.74 Å).

Another covalently bonded molecule is HF. The electronic configuration of the F atom is $1s^2\,2s^2\,2p^5$. Let us then consider the H atom, with its 1s orbital, as approaching along the z axis of the F atom, as sketched in Figure 3.4, where the p_x and p_y orbitals are deemed each to have two electrons.

A molecular orbital can be fashioned in either of two ways from the $2p_z$ orbital of F and the 1s orbital of H. The more stable σ orbital, depicted in Figure 3.4, has only one node, close to the F atom and arising from the node of

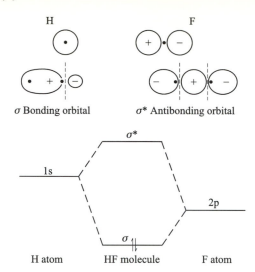

Figure 3.4   Sketch of the bonding (σ) and the antibonding (σ^*) orbitals in HF, and of the energies of these orbitals in relation to the 1s orbital of the H atom and the 2p orbital of the F atom.

the $2p_z$ orbital. The σ^* antibonding orbital has an additional nodal plane between the nuclei of H and F. Both of these orbitals have axial symmetry and so merit the designation 'σ' even though one of the basis orbitals is not an s orbital.

Also, in the case of HF, the energies of the basis orbitals are unequal, but that of the bonding MO lies below the level of the 1s orbital of the hydrogen atom by the same extent as the energy of the antibonding MO lies above that of the $2p_z$ orbital of the fluorine atom. In HF, the σ orbital is occupied by two electrons. In the hypothetical molecule, HeNe, there would additionally be two electrons occupying the σ^* orbital. The consequence would be that there would be no net bonding and it is for this reason that this molecule is non-existent.

The covalent bond in F_2 may be deemed to involve the $2p_z$ orbitals of the two atoms, with the nuclei aligned along their z axes. Once again, the resulting molecular orbitals both have axial symmetry. The σ or bonding MO has two nodes, one near each nucleus, while the σ^* antibonding MO has three nodes, of which one lies midway between the two nuclei. These are illustrated in Figure 3.5.

3.4 The pi bond in diatomic molecules

Under normal conditions, nitrogen exists as the strongly bonded diatomic molecule, N_2. Since N has the electronic configuration, $1s^2\ 2s^2\ 2p_x^{\ 1}\ 2p_y^{\ 1}\ 2p_z^{\ 1}$, all three of the 2p orbitals are capable of being involved in bonding.

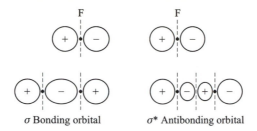

Figure 3.5 Sketch of the bonding (σ) and the antibonding (σ^*) orbitals in F$_2$.

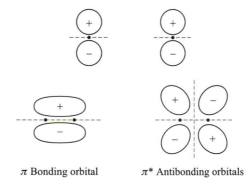

Figure 3.6 Illustration of the disposition of electron density in the π and the π^* orbitals formed by interaction of 2p orbitals at right angles to the internuclear axis.

Whereas the 1s orbitals are too small to be involved in bonding, the 2s and the 2p orbitals have comparable dimensions. The 2s orbitals of the two atoms interact to give σ_{2s} and σ_{2s}^* molecular orbitals. Both of these orbitals, the one bonding and the other antibonding, will be occupied by two electrons with spins paired. Consequently, the interaction of the 2s orbitals achieves no net bonding, and we must look towards the 2p orbitals to account for the strong bonding between the two N atoms.

If we consider the two N atoms as approaching along the z axis, then the $2p_z$ orbitals may interact as shown in Figure 3.5, to give the σ_{2p_z} and $\sigma_{2p_z}^*$ molecular orbitals. In addition molecular orbitals may also be formed by combination of the $2p_y$ orbitals of the two atoms. Since these are identical, they should have equal coefficients in the molecular orbitals, which will thus be $(\psi_a + \psi_b)$ and $(\psi_a - \psi_b)$, where the subscripts a and b distinguish the two atoms. The first of these molecular orbitals has one nodal plane through the internuclear axis and substantial overlap of electron density just above and just below the nodal plane, as is illustrated in Figure 3.6. This is a bonding molecular orbital, with a lower energy level than its basis atomic orbitals. It does not, of course, have axial symmetry, so it is denoted by the Greek letter π which corresponds to the Arabic letter p.

The second molecular orbital has two nodal planes, one through the internuclear axis and one perpendicular to it, midway between the two nuclei. With this additional node, a plane of zero electron density, the orbital is an antibonding one and is classified as π^*.

With the two N atoms aligned along the z axis, the respective $2p_x$ orbitals are positioned in just the same way as are the $2p_y$ orbitals, but in a different plane. So these atomic orbitals combine in just the same way as do the $2p_y$ orbitals, producing π and π^* molecular orbitals. Consequently, from the interaction of the 2p orbitals of the two N atoms, we have three bonding molecular orbitals, namely σ_{2p_z}, π_{2p_y} and π_{2p_x} and three antibonding molecular orbitals, $\sigma_{2p_z}^*$, $\pi_{2p_y}^*$ and $\pi_{2p_x}^*$. The relative energy levels of these are illustrated in Figure 3.7. The number of electrons to be accommodated is six, so in accordance with the Aufbau principle, these occupy the three bonding orbitals. In that sense, there is a triple bond in the molecule N_2, which would justify writing it as the structure, $N \equiv N$.

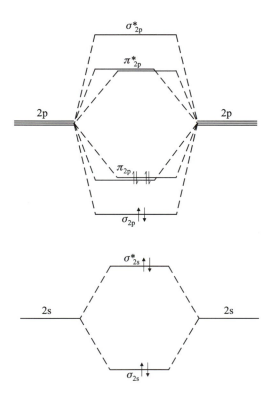

Figure 3.7 Representation of the energies of the bonding and the antibonding orbitals formed from the interaction of two atoms of second row elements. The occupancy of these reflects the structure of N_2.

However, of these bonds, two are similar, both being π bonds, but the other one, a σ bond, is different.

With NO, there is an additional electron. This can only be accommodated in one of the π^* orbitals, where it will tend to negate the bonding effect of the π orbitals, so that NO is not as strongly bonded as is N_2. The molecule O_2 has two electrons more than N_2. In allocating them, Hund's rule is relevant in view of the two π^* orbitals of equal energy. Consequently, one electron should go into the $\pi^*_{2p_y}$ and one to the $\pi^*_{2p_x}$ orbital, as indicated in Figure 3.8. This means that O_2 is more weakly bonded even than NO, and also that O_2, unlike N_2, will have two unpaired electrons. This is consistent with the known magnetic properties of O_2, which is paramagnetic, whereas N_2 and H_2 are diamagnetic. The molecule NO, with an odd number of electrons, must necessarily have one unpaired and is of course paramagnetic. H_2 and N_2 are diamagnetic and their MO structures do not imply that either should have any unpaired electrons.

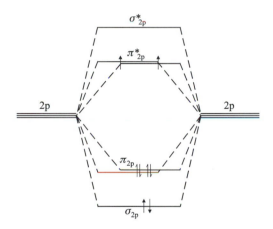

Figure 3.8 The same orbitals as are shown in Figure 3.7, reflecting the occupancy achieved in O_2.

3.5 Sigma bonding using hybrid atomic orbitals

In some polyatomic molecules, two or more monovalent atoms are attached to an atom of a different element with what are apparently identical bonds. In bonding terms, two issues then arise. One is whether, in order to explain the bonding in MO terms, it is possible to construct, from the atomic s and p orbitals of the central atom, bonding molecular orbitals which will be identical to each other. The other is whether the angles between these proposed molecular orbitals will be consistent with the known geometry of the polyatomic molecule.

A simple molecule of the type referred to above is beryllium hydride, BeH_2, which is linear with the two Be–H bonds having the same bond length. Beryllium has an atomic number of 4, and the electronic configuration $1s^2$ $2s^2$, which is not suggestive of any particular bonding scheme. If the configuration were $1s^2 2s^1 2p^1$, which is slightly higher in energy, this would suggest that two bonds could be formed, one using the 2s and one using a 2p orbital, but these bonds would not be identical.

The concept which solves this dilemma is that of orbital hybridisation and was first introduced by Pauling in 1931, in connection with his treatment of covalent bond formation in terms of the valence bond rather than the molecular orbital approach. In the present context it means that, qualitatively speaking, the available atomic orbitals are pooled and then divided up into equivalent portions. However, the process can be done with sufficient rigour to satisfy whatever tests may be applied to it. The hybrid orbitals have some s and some p character, in that each has a nodal plane, with unequal lobes of electron density to the two sides of the node. Of these two hybrid orbitals, denoted by the label sp, the large lobes point in opposite directions along the axis of the basis 2p orbital, as sketched in Figure 3.9.

If we regard the bonding of the two H atoms to Be on the basis that the latter has two sp hybrid orbitals, then it is apparent that each hybrid orbital, along with the 1s orbital of one of the H atoms, can give rise to a bonding MO of σ type. Each MO will closely resemble the σ orbital involved in the molecule HF, discussed in section 3.3, and each MO will be occupied by two electrons, of which one may be thought of as coming from the Be atom and one from the H atom. Thus the two Be–H bonds will be equivalent and, since the sp hybrid orbitals lie along the same axis, the two Be–H bonds will be at an angle of 180°.

One of the perennial questions about the validity of orbital hybridisation relates to the assumption that, for the Be atom, we may elevate an electron from the 2s to the 2p orbital to obtain the configuration $1s^2 2s^1 2p^1$. The answer is that this is, in effect, a self-financing venture in that the energy gap between the 2s and the 2p orbitals is less than the increase in the stability of the bonding orbitals that may be formed.

The 'pure' atomic orbitals

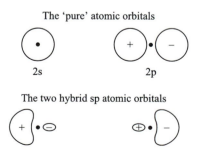

2s   2p

The two hybrid sp atomic orbitals

Figure 3.9 Illustration of the disposition of electron density in the sp hybrid orbitals.

Advancing to boron, of electronic configuration $1s^2 2s^2 2p^1$, this has to be treated, by analogy with beryllium, as if it had become $1s^2 2s^1 2p_y^1 2p_z^1$. There are now three orbitals that may go into the common pool, one of them an s orbital and two of type p, so the hybrid orbitals are labelled sp^2. (This, incidentally, is read 's-p-two'.) These all lie in the *yz* plane, at angles of 120° to each other, as shown in Figure 3.10. Clearly, a bonding molecular orbital, quite like the σ orbital in the molecule F_2, could be formed from a combination of each sp^2 hybrid orbital of the boron atom along with a 2p orbital of a fluorine atom. Consequently, the molecule BF_3 is planar, with each F–B–F angle equal to 120° and all B–F bonds equivalent.

Advancing one element further, to carbon, which has the configuration $1s^2 2s^2 2p^2$ and, similarly, needs to be treated as if it were $1s^2 2s^1 2p_x^1 2p_y^1 2p_z^1$, we now have the maximum number of atomic orbitals going into the common pool. The four resulting hybrid orbitals, designated as sp^3 (read as 's-p-three'), are arranged tetrahedrally, as sketched in Figure 3.11. Each sp^3 orbital has a shape very similar to that of the sp^2 hybrid orbitals described above. The sp^3 orbitals do not all lie in the same plane, but are disposed in three dimensions, with an angle of 109° 28′ between the axes of any two of these orbitals. Each of these hybrid atomic orbitals may, in concert with a 1s orbital of a hydrogen atom, form a σ molecular orbital, leading to the molecule CH_4, methane. As expected, in this molecule the H atoms are positioned tetrahedrally.

It is convenient to consider the bonding in the hydrides of the next two elements as variations on that of carbon. In bonding H atoms to N, we may conceive of the 2s and the three 2p atomic orbitals of the N atom being

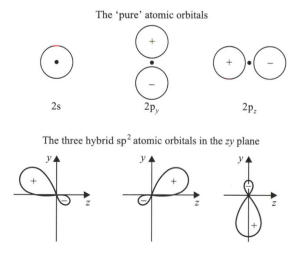

The 'pure' atomic orbitals

2s $2p_y$ $2p_z$

The three hybrid sp^2 atomic orbitals in the *zy* plane

Figure 3.10   Illustration of the electron density in the three sp^2 hybrid atomic orbitals.

The 'pure' atomic orbitals

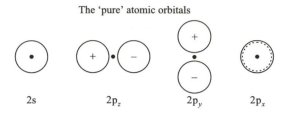

2s $2p_z$ $2p_y$ $2p_x$

The four hybrid sp^3 atomic orbitals arranged tetrahedrally

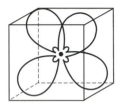

Figure 3.11   Illustration of the electron density in the four sp^3 hybrid atomic orbitals.

hybridised to give four equivalent sp^3 orbitals. Since N has five valence electrons, one more than carbon has, two of these electrons must share one of these hybrid orbitals. This leaves one electron for each of the remaining three hybrid orbitals and so three σ bonds may be formed between the N atom and three H atoms.

In this molecule, NH_3, the N atom has three σ bonds and a doubly-occupied hybrid atomic orbital, called a lone pair. The latter, with no positive nucleus attached, will repel each of the σ bonding electron pairs more strongly than they will repel each other. Consequently, the actual shape of this molecule will be that of a slightly distorted tetrahedron, with the σ bonding molecular orbitals a little closer to each other than they are in methane. Experimentally, the H–N–H bond angle is found to be 107°.

Similarly, in water, H_2O, the O atom has six valence electrons so that two of its sp^3 hybrid orbitals are occupied by lone pairs, leaving two to form σ bonds with H atoms. In terms of inter-orbital repulsions, there are now three categories. The mutual repulsion between two lone pairs is greater than that between a lone pair and a bonding pair, which in turn is greater than that between two bonding pairs. So water also is a slightly distorted tetrahedron. Experimentally, the H–O–H bond angle is found to be 104½°.

3.6   The pi bond in polyatomic systems

Among the compounds of carbon there are many containing more than one carbon atom. In ethane, C_2H_6, each carbon atom may be regarded as sp^3 hybridised, with one hybrid orbital from each atom serving as a basis orbital for the σ molecular orbital bonding the two carbon atoms. Each of the remaining sp³ hybrid orbitals is used in sigma bonding to an H atom, just as in methane. In this molecule, the bond between the two carbon atoms has axial symmetry and so is unaffected by the rotation of one methyl group with respect to the other. In practice, the internal rotation of C_2H_6 is not totally free, but this arises from the interaction of H atoms bonded to the separate C atoms rather than from purely bonding considerations.

In ethene, C_2H_4, each C atom is bonded to only two H atoms, with a double bond between the carbon atoms. In MO terms, each C atom is sp² hybridised. Let us assume that in each case, the 2s, the $2p_y$ and the $2p_z$ orbitals go into the pool, leading to three hybrid orbitals in the yz plane, and that the $2p_x$ orbitals, perpendicular to this plane, are left intact. These sp² hybrid orbitals, sketched in Figure 3.12, can form a σ bond between the two carbon atoms and can form

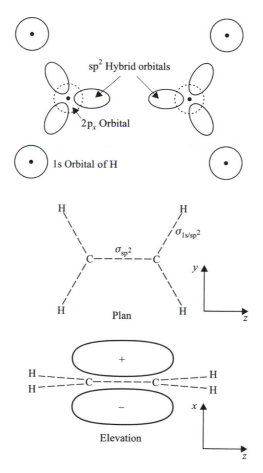

Figure 3.12   Illustration of the σ and the π bonds involved in ethene.

σ bonds between each C atom and two H atoms. The $2p_x$ orbitals, projecting above and below this plane, can form a π bond similar to those described above in regard to the molecule N_2.

So, in ethene, the two bonds between the C atoms are of different types, one σ and one π. The latter is not axially symmetric and, in order that there can be good overlap of electron density between the two basis orbitals, these must lie in the same plane through the internuclear axis. Consequently, the six atoms in C_2H_4 should all be coplanar. Experimentally this is found to be the case and the H–C–H angle is 118°, which is very close to the value of 120° predicted on the basis of sp^2 hybridisation.

For ethyne, C_2H_2, one expects sp hybridisation, using the 2s and the $2p_z$ orbitals and leaving intact the $2p_x$ and the $2p_y$ atomic orbitals. The resulting sp hybrid orbitals lie along the z asis. On each carbon atom, one sp hybrid orbital is involved in a σ bond with the other carbon atom and the second forms a σ bond with a hydrogen atom, as shown in Figure 3.13.

The $2p_x$ and $2p_y$ orbitals form two π bonds, one in the vertical and one in the horizontal plane through the internuclear axis of the two C atoms. Thus, as in N_2, there is a triple bond between the two C atoms, consisting of a σ and two π bonds. This interpretation of the bonding in ethyne implies that this molecule should be linear, as it is known to be.

The usefulness of these ideas of σ bonding using hybridised orbitals along with π bonding using pure p orbitals is also demonstrated by allene, $H_2C:C:CH_2$. The central atom forms two double bonds, so it must be sp hybridised. We assume this leaves the $2p_x$ and $2p_y$ orbitals intact. The terminal carbon atoms resemble those in ethene, and so are sp^2 hybridised. Let us assume that in the left-hand carbon atom, the $2p_y$ orbital is intact, so that the sp^2 hybrid orbitals, and the H atoms to which two of these form σ bonds, lie in the horizontal plane. The π bond between the left-hand and the central C atoms is then formed from the $2p_y$ orbitals and lies in the vertical plane, as shown in Figure 3.14.

The central and the right-hand carbon atoms are σ bonded and π bonded in like fashion, but this π bond must use the $2p_x$ orbitals and lie in the horizontal plane. Consequently, the sp^2 hybrid orbitals of the right-hand carbon atom, and the H atoms to which two of them are bonded, must lie in the vertical plane, as shown in Figure 3.14. This prediction, that the four H atoms of allene

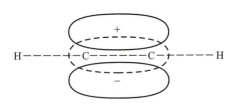

Figure 3.13   Illustration of the σ and the π bonds involved in ethyne.

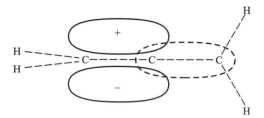

Figure 3.14   Illustration of the σ and the π bonds involved in allene.

do not all lie in the same plane, is known to be correct: experiment shows that the three carbon atoms are collinear, and that the planes of the two pairs of H atoms lie at 90° to each other.

In organic chemistry, a very important group in terms of its reactivity and its spectroscopic behaviour is the carbonyl group, $>C=O$. As our last example, let us consider methanal (formaldehyde), H_2CO. The carbon atom, bonded to three different atoms, is presumed to be sp^2 hybridised, leaving the $2p_y$ orbitals intact. Two of these hybrid orbitals form σ bonds with the H atoms. The oxygen atom is assumed to be in the state $1s^2\ 2s^2\ 2p_x^2\ 2p_y^1\ 2p_z^1$, and a σ bond is formed between C and O using the third hybrid orbital of the former and the $2p_z$ orbital of the latter. In addition, a π bond is formed between these same atoms using their $2p_y$ orbitals.

3.7   Delocalised molecular orbitals and aromatic systems

The extent and significance of organic chemistry is very much a consequence of the high valency of carbon, which is four, and of the ability of carbon atoms to bond with each other. To date we have encountered C–C single bonds, C=C double bonds and C≡C triple bonds, and there is a progressive shortening of the internuclear distance along this series. However, although in some molecules there seems to be a clear distinction between the long C–C single bonds (about 1.54 Å) and the shorter double bonds (about 1.34 Å), in others this is more blurred.

The unsaturated hydrocarbon, 1,3-butadiene is usually written as $CH_2:CH.CH:CH_2$, with a single bond between two double bonds. However, although the C–C single bond in propene, $CH_3.CH:CH_2$, is of the normal length, the bond length of the central single bond in butadiene is only 1.48 Å. Another feature is that the UV spectrum of 1,3-butadiene differs appreciably from that of an alkene. The question which arises is whether, in a molecule of this type, the bonding is adequately described in terms of two two-centre π bonds.

In applying molecular orbital theory to 1,3-butadiene, each carbon atom is considered to be sp^2 hybridised. If, as they are sketched in Figure 3.15, these hybrid orbitals and the resulting σ bonds are all assumed to be in the same plane, then there is the maximum opportunity for interaction of the four remaining 2p orbitals. The number of molecular orbitals that ensues must also be equal to four. The molecular orbital of lowest energy is the one obtained by positive combination of all four of the atomic orbitals, with the only node being in the plane through the four nuclei. The second most favourable orbital has a second nodal plane between the two central carbon atoms, as indicated in Figure 3.15. Each of these delocalised molecular orbitals accommodates two electrons and both have lower energies than the localised orbital of an alkene. The other two MOs have higher energies and are antibonding.

The molecule which, most of all, exemplifies the role of delocalised molecular orbitals is benzene, C_6H_6. The elucidation of the structure of this molecule posed some problems. When its molecular formula was known, it

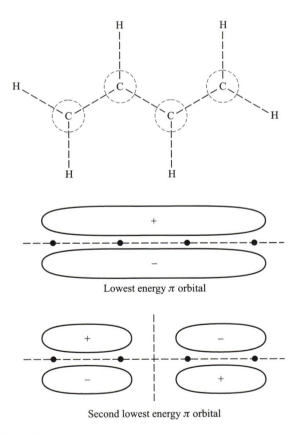

Figure 3.15 Illustration of the hybrid atomic and the delocalised π orbitals in 1,3-butadiene.

seemed, from its ratio of H atoms to C atoms, to be a very 'unsaturated' molecule, which ought to be extremely reactive. Kekulé asserted that it was cyclic, and that its structure could be represented in terms of an oscillation between two structures, each with three double bonds, arranged uniformly around the ring:

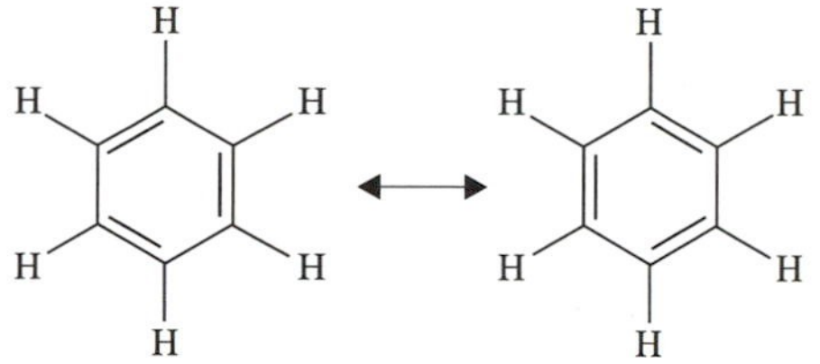

From a chemical standpoint, one anomaly was that whereas alkenes react readily with bromine, benzene does not. The ultraviolet spectrum of benzene shows intense absorption, with a maximum intensity at 255 nm, whereas alkenes absorb very little except in the vacuum UV, not accessible using a normal spectrophotometer. Moreover, the nuclear magnetic resonance spectrum of benzene is totally different from that of an alkene. Also, all C–C bond lengths in benzene are equal and, at 1.40 Å, they lie between those of normal single and double bonds.

The molecular orbital approach to benzene follows logically from our treatment of 1,3-butadiene. If the bonding orbitals of each C atom are sp^2 hybridised, the angle between these hybrid orbitals is 120°, which is the correct internal angle for a regular hexagon. Thus we may have all six C atoms and their hybrid orbitals in the same plane, with each carbon atom σ bonded to its two neighbours and to one H atom. This leaves intact the other 2p orbital of each atom, perpendicular to the plane of the ring. Molecular orbital calculations show that the most stable orbital available is that with equal contributions from all six basis atomic orbitals, with just one node, lying in the plane of the ring. Next, there come two other orbitals, illustrated in Figure 3.16, each with one additional nodal plane, but both are lower in energy than a localised two-centre π orbital, as in an alkene. So, two electrons are allocated to each of these three delocalised, bonding, π molecular orbitals, and the other three molecular orbitals, all antibonding, are unoccupied.

Thus in benzene, where the π electrons are totally delocalised, the best representation of the structure is perhaps as follows,

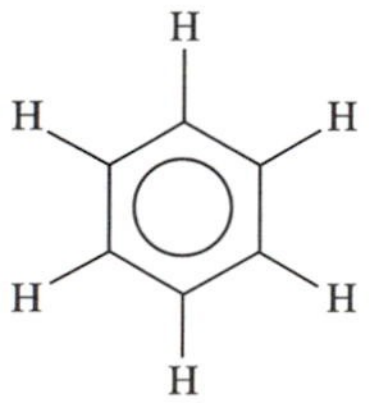

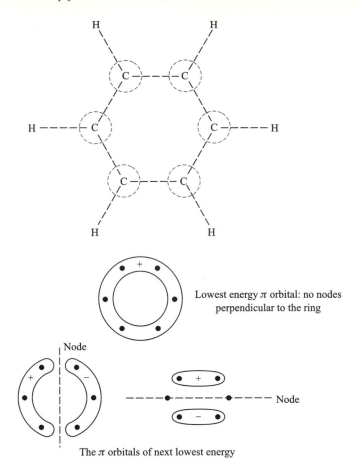

Figure 3.16 Illustration of the localised σ bonds and the delocalised π orbitals in benzene.

rather than the Kekulé structures shown above. This makes clear that, while benzene is an 'unsaturated' hydrocarbon, the bonding in it is quite different from that in an alkene. The symbol is also suggestive of the ease with which, in a molecule containing a benzene ring, electrical effects are conveyed from one end to the other.

More extensive analysis of systems of this nature shows that, in order to have total delocalisation of the π electrons in what is termed an **aromatic** system, it is necessary that the number of electrons involved should belong to the series, 6, 10, 14, 18, ... Compounds with fused benzene rings, like naphthalene (10 π electrons) and anthracene (14 π electrons), all meet this requirement and are aromatic compounds.

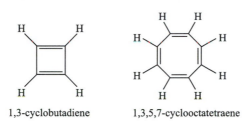

Naphthalene   Anthracene

However, 1,3-cyclobutadiene and 1,3,5,7-cyclooctatetraene, while they may be represented by structures with alternating single and double bonds (which is a feature of the Kekulé structures of benzene) are not aromatic compounds. In MO terminology, having four and eight π electrons, they do not meet the necessary criterion. In fact, 1,3-cyclobutadiene has never been synthesised and 1,3,5,7-cyclooctatetraene is non-planar and behaves as an alkene.

1,3-cyclobutadiene   1,3,5,7-cyclooctatetraene

3.8 Co-ordination complexes of transition metals

The cations of the transition metals are capable of forming chemical bonds with a range of molecules and anions. In these, the metal ion occupies a central position with the set of molecules and ions, collectively known as ligands, attached around it. The number of ligands around an ion may have any value between two and nine, but more frequently it is four or six.

One common feature of potential ligands is that they have a lone pair of electrons which are used in bonding to the cation. Some are neutral molecules like NH_3 or CO or H_2O, some are simple anions like Cl^- or NO_3^- or CN^-, while others are more complex organic molecules or ions.

The metal ion usually carries a charge of at least $2+$, so that its outer s electrons (4s in the case of the first transition series) have been shed. For bonding purposes, its most important orbitals are then of type d: for example, for the ion Co^{3+}, there are six electrons in the 3d orbitals.

The details of the bonding of the ligands to the cation are of considerably greater complexity than those discussed in the previous sections. Only a much simplified treatment of what is called ligand field theory will be offered here.

53

Let us start by supposing that we have six ligands disposed around a transition metal ion, with two on each of the x, y and z axes. As each ligand approaches the central metal ion, its lone pair is presumed to be directed along the metal–ligand axis, in which case it has the same axial symmetry as a σ orbital. In the interaction of these six ligand orbitals, only certain orbital combinations are allowed and these are of the same symmetry as two of the 3d orbitals of the cation (the $3d_{x^2-y^2}$ and the $3d_{z^2}$) but different from the other three 3d orbitals. Consequently, the bonding molecular orbitals between the ligands and the metal are constructed from only two of the five 3d orbitals, and the other three remain as non-bonding orbitals.

The result of the allocation of electrons to the available orbitals, bonding, non-bonding, and antibonding, depends on how many electrons are being supplied by the metal ion. If six, then these, along with the 12 electrons from the six ligands, will occupy the six bonding and the three non-bonding orbitals. In this way, an ion like Co^{3+} can form a complex such as $[Co(CN)_6]^{3-}$, in which each CN^- ligand is attached to the metal ion by a bond involving two electrons.

Metal–ligand bonding tends to be even stronger for the transition elements of the second and especially the third transition series, as their 4d and 5d orbitals, which are less compact than the 3d orbitals, have more electron density protruding far from the nucleus. Also, some ligands bond more strongly than others, the anion CN^- and the molecule CO being two of the stronger ones. The poisonous nature of cyanide and carbon monoxide arises from their strong tendencies to bind to the iron ions in haemoglobin or cytochrome c.

Suggested reading

MURRELL, J. N., KETTLE, S. F. A. and TEDDER, J. M., 1985, *The Chemical Bond*, 2nd Edn., Chichester: Wiley.

KARPLUS, M. and PORTER, R. N., 1970, *Atoms and Molecules: An Introduction for Students of Physical Chemistry*, New York: Benjamin.

WEBSTER, B. C., 1990, *Chemical Bonding Theory*, Oxford: Blackwell.

STREITWIESER, A., 1961, *Molecular Orbital Theory for Organic Chemists*. New York: Wiley.

DeKOCK, R. L. and GRAY, H. B., 1980, *Chemical Structure and Bonding*, Menlo Park, CA: Benjamin/Cummings.

Problems

3.1 Specify the presumed state of atomic orbital hybridisation of each of the carbon atoms in the following compounds:

(a) ethanal, CH_3CHO
(b) 2-pentyne, $CH_3C{\equiv}CCH_2CH_3$

(c)   1,1-difluoropropene, $F_2C=CHCH_3$

In the case of compound (b), what is the maximum number of collinear atoms?

3.2   Some divalent metals react with O_2 to form a peroxide, MO_2, containing the ion, O_2^{2-}, whereas some univalent metals react to form a superoxide, of similar molecular formula, but containing the ion, O_2^{-}. On the basis that the presence of unpaired electrons is a symptom of paramagnetism, would you, on molecular orbital grounds, expect (a) a metal peroxide, MO_2 or (b) a metal superoxide, MO_2, to be diamagnetic or paramagnetic?

4

Introduction to chemical spectroscopy

One of the adages of preparative chemistry is that when someone has prepared a new compound, the first priority is to run its spectrum. Three major spectroscopic techniques are introduced in this chapter: infrared (IR) spectroscopy, which documents the vibrational modes of a molecule and reflects its functional groups; ultraviolet (UV) spectroscopy, which detects transitions of the bonding electrons to higher electronic states, indicative of the electronic structure; and nuclear magnetic resonance (NMR) spectroscopy, which detects transitions in the spin states of the nuclei, of such a nature that the absorptions observed convey even more structural information than do the IR and UV techniques. With NMR, structural questions which formerly were resolved only after patient endeavours and great ingenuity can now confidently be answered in an afternoon.

4.1 The electromagnetic spectrum

The various types of electromagnetic radiation all involve mutually perpendicular oscillating electric and magnetic fields, in the plane perpendicular to the direction in which the radiation is travelling. In a vacuum, they all have the same very considerable velocity c, equal to 2.998×10^8 m s^{-1}.

Travelling waves have two significant characteristics. One is the wavelength, which is the distance between corresponding points on adjacent waves, and is usually denoted by the symbol, λ. The other is the frequency, ν, which may be defined as the number of wave crests to pass a stationary point in unit time. The product of these two quantities is the velocity of the wave:

$$c = \lambda \nu \tag{4.1}$$

For electromagnetic radiation of all types, the velocity is the same. Thus radiation of long wavelength has a low frequency, whereas short wavelength radiation has a high frequency. The various types of radiation in the electromagnetic spectrum are shown in Figure 4.1, in terms both of wavelength and of frequency.

Light exhibits a curious duality. While it has a wave nature, as illustrated by its diffraction, it consists of discrete packets, called quanta. By Planck's formula, the energy, ε, of a quantum is directly proportional to the frequency, v, and is expressed by the equation,

$$\varepsilon = hv \tag{4.2}$$

where h stands for the fundamental constant named after Planck.

This means that electromagnetic radiation of long wavelengths, and thus low frequencies, comes in quanta of very low energy. For example, if we consider the wavelength, in the radio range, of 300 m, then the corresponding frequency is given by,

$$v = \frac{c}{\lambda} = \frac{3.0 \times 10^8 \text{m s}^{-1}}{300 \text{ m}}$$
$$= 10^6 \text{ s}^{-1}$$

and the energy of a quantum may be evaluated:

$$\varepsilon = hv = 6.63 \times 10^{-34} \text{ Js} \times 10^6 \text{ s}^{-1}$$
$$= 6.63 \times 10^{-28} \text{ J}$$
$$\equiv 4.14 \times 10^{-9} \text{ eV}$$

We may contrast this with the energy of a quantum of X-rays. If the wavelength were 1.5 Å which is 1.5×10^{-10} m, then the corresponding frequency is $2.0 \times 10^{18} \text{ s}^{-1}$. This gives, for the energy of the quantum of X-rays,

$$\varepsilon = 6.63 \times 10^{-34} \text{ Js} \times 2.0 \times 10^{18} \text{ s}^{-1}$$
$$= 1.33 \times 10^{-15} \text{ J}$$
$$\equiv 8.3 \times 10^3 \text{ eV}$$

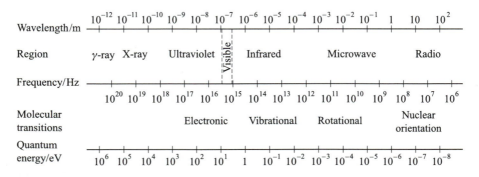

Figure 4.1 The electromagnetic spectrum.

As Figure 4.1 shows, visible light lies between these two types of radiation and it consists of quanta of a very few electron-volts energy.

While it is convenient to characterise light by its wavelength, as these calculations illustrate, the relationship between wavelength and quantum energy is an inverse one. The reciprocal of the wavelength is much more indicative of the energy involved, and spectroscopists frequently use such a quantity. This habit pre-dates the acceptance of SI units, so the figure quoted is the number of wavelengths per cm. This parameter, which we shall denote by the wavenumber, ω, needs to be multiplied by the velocity of light, in cm s^{-1}, to convert it into the frequency, v. The positions of absorption peaks in the infrared are usually cited as so many wavenumbers, e.g. 1710 cm^{-1}, and the energy of the IR quantum involved is directly proportional to this figure:

$$\varepsilon = hv$$

$$= 6.63 \times 10^{-34} \text{ Js} \times 3.0 \times 10^{10} \text{ cm s}^{-1} \times 1710 \text{ cm}^{-1}$$

$$= 3.40 \times 10^{-20} \text{ J}$$

$$\equiv 0.212 \text{ eV}$$

4.2 Infrared spectroscopy: the theory

Changes to the internal energy of a large molecule may, to an approximation, be subdivided into those involving changes in the energy of rotation, or of vibration, or to changes in its electronic state. While a macroscopic body like a top may spin at any speed, fast or slow, the rotation of a molecule is quantised, so that only certain speeds of rotation are possible. The difference in the energies of adjacent rotational states is fairly small, so that the quantum of energy equal to this difference usually belongs to the microwave range.

Microwave studies of gas phase molecules can be extremely useful for measuring the internuclear separations within the molecule. However, this type of spectroscopy is rather specialised and is not readily applicable to high molar mass compounds of low volatility, so it will not be further discussed here.

The vibration of a molecule is also quantised, with only certain amplitudes of vibration being possible. This may be illustrated by reference to the hypothetical diatomic molecule, AB. To a fairly good approximation, the dependence of the potential energy of such a molecule on the internuclear separation is described by a parabola, as shown in Figure 4.2. In this situation, the permitted vibrational energies are given by,

$$\varepsilon = (V + \tfrac{1}{2})\frac{h}{2\pi}\sqrt{\frac{k}{\mu}} \qquad (4.3)$$

where the vibrational quantum number V may be zero or a positive integer, h is Planck's constant, k is the force constant, which is a measure of the rigidity of the bond, and μ denotes the reduced mass of the AB molecule, given by $\mu = m_A m_B/(m_A + m_B)$. These are illustrated in Figure 4.2.

An important point about the vibration of a molecule is that both nuclei will move, with the centre of gravity remaining static. It is for this reason that the relevant equation, (4.3), contains the reduced mass. The lighter of the two nuclei will have a greater amplitude to its motion than will the heavier one. So, in a molecule like HI, nearly all the vibrating is done by the H atom, while the I atom remains almost stationary.

As Equation (4.3) indicates, the ground vibrational state, with $V = 0$, corresponds not to a static molecule, but to one in which some vibrational motion persists. This may be seen as one manifestation of Heisenberg's Uncertainty Principle. The amount by which the ground vibrational state lies above the base of the potential energy curve is given by,

$$\varepsilon = \frac{h}{4\pi}\sqrt{\frac{k}{\mu}} \tag{4.4}$$

and is called the zero-point energy. This depends, inversely, on the square root of the reduced mass: thus the zero-point energy of the light molecule, H_2, is very much greater than that of the far heavier I_2.

At ambient temperatures, only the ground vibrational state is appreciably populated, so vibrational spectroscopy is usually studied in absorption. The permissible change in V is that V may increase by 1, but only if the vibration causes a change in the dipole moment of the molecule. Thus in the symmetrical

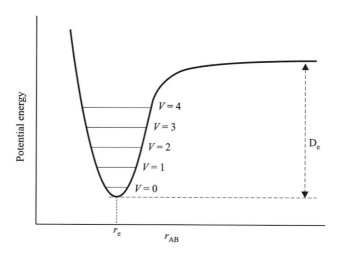

Figure 4.2 Illustration of the energies of the permitted vibrational energy levels of the diatomic molecule, AB, from the plot of potential energy as a function of the internuclear separation, r_{AB}. The bond dissociation energy of the AB molecule is represented by the parameter, D_e.

molecule N_2, vibrational change is excluded and it is IR inactive, whereas the unsymmetrical molecule CO is IR active. However, in large polyatomic molecules, most vibrations are IR active.

Whereas a diatomic molecule has only one vibration, of the bond stretching variety, polyatomic molecules have both stretching and bending vibrations. Molecules usually have more rigidity in respect of bond stretching than of bond bending, so the latter category of vibration tends to have a lower value of the force constant, k. From Equation (4.3) it may be seen that bending vibrational states will lie closer together than do stretching vibrations, so that the transitions will involve less energy. A non-linear molecule of n atoms has, in total, $(3n-6)$ vibrations. This means that ethanal, CH_3CHO, for example, has 15 vibrational modes.

4.3 Infrared spectroscopy: experimental and interpretation

A simplified diagram of an IR spectrometer is shown in Figure 4.3. 'White' radiation in the IR is emitted by a Nernst source, composed of oxides of the rare earths and heated to about 1700°C. This passes through the sample and then, via collimating mirrors, it is dispersed by a prism, whose angle determines which wavelength of IR radiation reaches the back slit and thus the detector. By having the rotation of the prism synchronised with that of the chart recorder, the spectrum may be recorded on chart paper over the accessible range.

Rocksalt, NaCl, is frequently used as the material of the prism and for the plates holding the sample in position. It transmits between 4000 and 667 cm^{-1}, whereas CsI, which is much more expensive, has a far larger range of transmission and permits measurements to be made down to 185 cm^{-1}. Both these materials are extremely water-soluble, so the prism and the plates must be kept dry at all times.

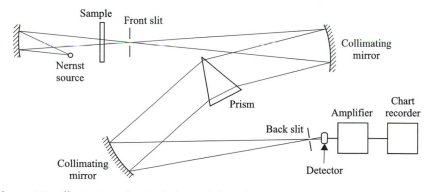

Figure 4.3 Illustration of a single-beam infrared spectrometer.

On the chart paper used with an IR spectrometer, the abscissa is usually graduated in cm^{-1}, and the ordinate in percentage transmittance. This means that, in the absence of an absorbing sample, a horizontal line is obtained along the top of the chart. If there is a sample with a few absorption bands, then the horizontal line shows occasional dips towards lower transmittance, as is exemplified in Figure 4.4.

One of the simpler types of vibration to rationalise is a stretching vibration involving an atom bonded only to one other atom, such as C–H. While all such systems have virtually the same reduced mass, the rigidity of the C–H bond, and thus the vibration frequency, is influenced by the chemical environment of the carbon atom. In Figure 4.4, there are absorption bands centred at 3050, 2810 and 2725 cm^{-1}. These are all due to C–H stretching vibrations, the first one arising from the aromatic hydrogens of the benzene ring and the others from the aldehydic hydrogen. If D atoms were to be substituted for H in either group, the consequence for the IR absorption band would be considerable. The reason is that this would almost double the reduced mass of the vibration, so that, to a fair approximation, the absorption frequency would be decreased by a factor of $\sqrt{2}$.

Another readily interpreted vibration is the C=O stretching mode. The reduced mass is very much greater than for C–H, but the double bond is more rigid than the single bond, so that, compared with C–H, both μ and k have increased, but the former by a greater factor. The absorption band in Figure 4.4 at 1695 cm^{-1} is attributable to this vibration. While the exact position of a C=O stretching band is influenced by the chemical environment, if the IR spectrum of a substance shows no peak between 1600 and 1750 cm^{-1}, this must rule out any possibility that the compound is an aldehyde, a ketone, a carboxylic acid, an ester or an amide.

In Figure 4.4, there are several intense absorptions between 1600 and 800 cm^{-1}. These are characteristic of the benzene ring, and arise from various vibrations of the ring itself. The positions of these bands vary depending on

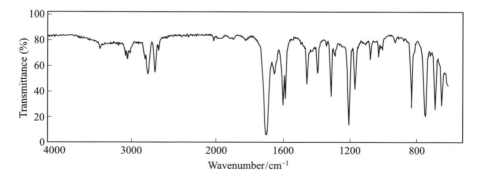

Figure 4.4   An infrared spectrum of benzaldehyde, C$_6$H$_5$.CHO, in the form of a plot of the transmittance, as a percentage, against ω, in cm^{-1}.

whether the ring is unsubstituted or has one substituent, or has two substituents in the 1- and 2-, or in the 1- and 3-, or in the 1- and 4- positions, or is more extensively substituted. The band at around 750 cm^{-1} is attributable to a bending mode of the aromatic C–H bond.

Table 4.1 lists the positions at which the most common vibrational modes may be detected. These are all to be read as stretching vibrations unless otherwise stated. On such a basis, intelligent analysis of the IR spectrum of a compound may enable its identity to be established. Where doubt persists, it is usually profitable to compare the spectrum with that of an authentic sample of the suspected compound. The extent of the agreement between the two is definitive.

4.4 Ultraviolet/visible spectroscopy: experimental aspects

A modern ultraviolet spectrophotometer has an operational range of perhaps 185 to 900 (or 1000) nm. While only a minority of this wavelength range lies in

Table 4.1 Location of some vibrational peaks exhibited by organic compounds

Vibration	ω/cm^{-1}
O–H	3600
N–H	3500
C–H (alkyne)	3300
C–H (alkene)	3050
Ph–H	3050
C–H (alkyl)	2850–3000
S–H	2580
C≡N	2250
C≡C	2220
C=O ⎰ aldehyde / ketone / carboxylic acid / ester ⎱	1700–1750
C=O (amide)	1640–1700
C=C (alkene)	1650
–NO$_2$	{ 1500 / 1350
–CH$_2$ – (scissors mode)	1465
C–O (ester)	1250
C=S	1100
C–O (ether)	1100
C–F	1050
Ph–H (bending)	760

Ph denotes the phenyl group, C_6H_5, so that Ph–H denotes a C–H bond in benzene.

the UV, if one were to convert to a scale which is linear in frequency, then this range would appear to lie predominantly in the UV. The lower wavelength limit is imposed by the fact that below 185 nm, O_2 (dioxygen) absorbs appreciably and so air needs totally to be excluded: hence the region below this limit is labelled the vacuum UV and it requires special apparatus.

In an ultraviolet spectrophotometer, there are two lamps, as alternative light sources, a deuterium lamp for use over the range 185–360 nm and a tungsten or a quartz–halogen lamp for use over the remainder. When switching lamps, the first mirror is rotated so that the light from the new lamp is directed through the slit. To obtain a monochromatic beam, an interference grating is almost invariably used, after which the beam is split into two, with one half passing through the reference cell and the other through the sample cell, as illustrated in Figure 4.5. When operating, the lid of the cell compartment is kept closed so as to exclude stray light.

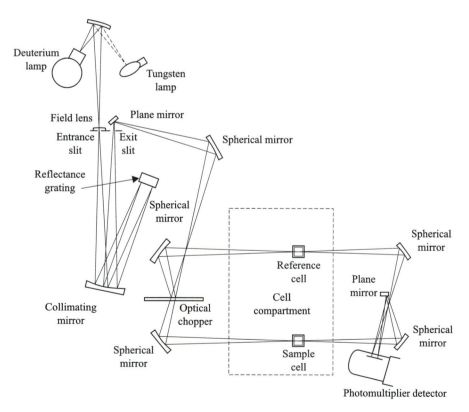

Figure 4.5   Illustration of a double-beam UV/visible spectrophotometer.

Most UV spectroscopy is carried out using solutions and employing a transparent solvent. The cells containing the solution must also freely transmit the light employed: quartz cells are normally used but the much cheaper pyrex cells would suffice if measurements were to be made only at wavelengths in excess of 365 nm. Photomultiplier tubes, with two different types to cover the wavelength range, are used as detectors, operating under conditions where the photocurrent is strictly proportional to the intensity of the incident light. Thus the ratios of the photocurrents produced by the beams passing through the sample and the reference cells can be measured. In a double beam instrument, such as that illustrated in Figure 4.5, this measurement will be unaffected by fluctuations in the output of the lamp.

As mentioned above, the solvent should be totally transparent over the region scanned. The significance of the reference cell is then that it will cause the same fraction of the light to be reflected from its walls as will occur with the sample cell. Thus the ratio of the intensities of the light transmitted through the sample and the reference cells will be totally attributable to the solute.

Suppose a beam of monochromatic light, of intensity I_o, passes into the solution in a cell, of path length l cm, and that the intensity of the transmitted light is I_t. The extent of the absorption can be expressed by the transmittance, I_t/I_o, as is usually done in IR spectroscopy. Where any absorption occurs, the transmittance will be less than unity. There is, however, an advantage in reporting the absorbance, A, defined as,

$$A = \log_{10}(I_o/I_t) \tag{4.5}$$

for the reason that A (which is dimensionless) is proportional to the path length, l, of the cell and (almost invariably) to the concentration, c, of the solution. This leads to the equation,

$$A = \varepsilon c l \tag{4.6}$$

which is a statement of the Beer–Lambert Law. The concentration, c, is usually given in the units, mol dm^{-3}, so that ε, the molar decadic absorption coefficient, is in the units, dm^3 mol^{-1} cm^{-1}.

In running a UV spectrum, one normally uses, as the ordinate, the absorbance, A. Since $A = 2.5$ represents the transmission of only 0.3% of the incident light, absorbance values of this magnitude need to be treated with caution, as a minute amount of stray light can lead to a very large error in A. A more accurate spectrum will usually be obtained by using a shorter path length cell or a lower concentration of solute.

A UV spectrum of benzene in *n*-hexane is shown in Figure 4.6. This demonstrates that, within this range, the wavelength of maximum absorption, frequently called λ_{max}, is 255 nm. Using a cell of 1-cm path length, the absorbance at this wavelength attains the value of 0.90, with the benzene concentration equal to 5×10^{-3} mol dm^{-3}.

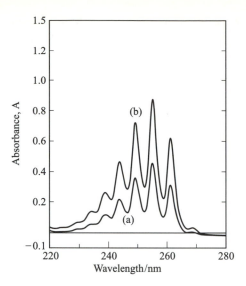

Figure 4.6 A UV spectrum of benzene in *n*-hexane, at concentrations of (a) 2.5 × 10^{-3} and (b) 5 × 10^{-3} mol dm^{-3}, in a cell of 1-cm path length.

Thus at this wavelength we have:

$$\varepsilon = \frac{A}{cl}$$

$$= \frac{0.90}{5 \times 10^{-3} \text{ mol dm}^{-3} \times 1 \text{ cm}}$$

$$= 1.8 \times 10^2 \text{ dm}^3 \text{ mol}^{-1} \text{ cm}^{-1}$$

4.5 The UV spectra of organic compounds

Light absorption within the range of a UV spectrophotometer occurs because of electronic transitions where the energy difference between the two states lies in the range of a few electron-volts. In some substances, such as alkanes, the only possible electronic excitations would require a much greater energy, so these compounds are totally transparent throughout this range of wavelengths.

In an alkane, there are core electrons and electrons engaging in σ bonding, either between two C atoms or between C and H. As has been detailed in Chapter 3, where a σ molecular orbital is formed by the combination of two atomic orbitals, it is also possible to form an antibonding σ* orbital. The transition by which an electron is elevated from a σ to a σ* orbital (and written, σ → σ*) will require so much energy that the appropriate quantum will belong to the vacuum UV. In an alkene, there is also a π bond between two C atoms, and the corresponding transition, $\pi \to \pi^*$, is feasible. This energy difference is not quite so large but once again it corresponds to the vacuum UV region.

In Chapter 3, the nature of 'conjugated' double bonds was explored, using butadiene as an example. This compound has partially delocalised molecular orbitals and the gap between the highest occupied π orbital and the lowest unoccupied π^* orbital is appreciably less than in an alkene. So 1,3-butadiene shows an absorption peak at 215 nm. Benzene has totally delocalised π orbitals and the corresponding gap is even smaller, so it shows substantial absorption in the UV with λ_{max} at 255 nm, as shown in Figure 4.6. In more extensive aromatic systems, the $\pi \to \pi^*$ absorption bands extend to even longer wavelengths.

In a carbonyl group, the O atom is bonded to the C atom by both a σ and a π bond. The energy gap between the π and the π^* orbitals is of such a magnitude that the $\pi \to \pi^*$ band of, for example, 2-propanone lies in the vacuum UV at 166 nm. Where the carbonyl group is adjacent to a π electron system, the interaction of the two causes the energy gap in the former to be decreased. For example, in acetophenone, $Ph.CO.CH_3$, the $\pi \to \pi^*$ absorption band is found at 238 nm and is fairly intense with $\varepsilon = 1.34 \times 10^4$ dm^3 mol^{-1} cm^{-1}.

Carbonyl compounds also show another and much weaker band at longer wavelengths. This arises because of the excitation, not of a bonding electron, but of an electron from a non-bonding 2p orbital on the O atom. Since the energy level of a bonding π orbital lies below that of a 2p orbital, it follows that less energy can raise an electron from the 2p to the π^* orbital than is needed for the $\pi \to \pi^*$ excitation. The shorthand label for this process is $n \to \pi^*$.

Earlier in this chapter it has been implied that not all conceivable transitions from one state to another may actually occur. Within electronic spectroscopy, some transitions are 'allowed' and some 'forbidden'. In the case of the latter this means, in non-technical language, that there is an acute difficulty for the molecule in the lower electronic state to absorb a quantum of the correct size and attain the upper electronic state. This may arise because of the relative disposition of the two orbitals in space. In practice, a 'forbidden' transition can still take place, but it is much less probable than an 'allowed' transition, and will have a far lower value of ε. The $n \to \pi^*$ transition falls into this category and in 2-propanone the peak is at 272 nm with $\varepsilon = 14$ dm^3 mol^{-1} cm^{-1}, which is several orders of magnitude less than for the 'allowed' $\pi \to \pi^*$ transition.

4.6 The UV spectra of transition metal ions

Among inorganic cations in solution, two groups are distinguishable. There are those like the alkali metals (Group I) and the alkaline earths (Group II), which are invariably colourless and transparent in the UV. Then there are those of the transition metals, which are nearly all coloured. The most obvious difference between ions such as Ca^{2+} and Cs^+ on the one hand and Cu^{2+} or Ti^{3+} on the other is that the former have either no d electrons or a complete shell of d electrons while the latter have an incomplete shell of d electrons.

In an isolated atom, the five 3d orbitals all have the same energy. However, when a cation is in a typical chemical environment, it is not isolated. In solution, polar molecules of solvent will interact with it and will become attached around the ion as ligands. If a species is present which is a good electron donor, such as the CN^- anion, the interaction will be all the greater. The presence of the ligand creates an electric field around the ion, under whose influence the energies of the 3d orbitals are split. The effect of this phenomenon depends on the number and the geometrical arrangement of the ligands.

Where there are six ligands, arranged octahedrally, the energies of the d_{xy}, d_{yz} and d_{xz} orbitals are lowered and those of the $d_{x^2-y^2}$ and d_{z^2} are raised. The energy difference between these two groups is usually denoted by Δ. An electronic transition between these is formally 'forbidden', but for technical reasons it may belong to the 'weakly allowed' category, so that d–d transitions may have low or moderate values of ε.

So d–d transitions may be detectable from metal ions with a partial shell of d electrons. This category does not, of course, include the ions Zn^{2+}, Cd^{2+} or Cu^+, which all have the configuration, d^{10}. On the other hand, the ion Cu^{2+} belongs to this group since it has the configuration, $3d^9$. The nature of the ligand influences not only the magnitude of Δ, and thus the wavelength of the d–d absorption maximum, but also the nature of the spectroscopic transition and consequently the intensity of the absorption. The size of Δ is such that it usually corresponds to a quantum within the visible range; consequently, transition metals are seen as having coloured compounds.

UV spectrophotometry is widely used for quantitative estimation in solution. In some cases, the species to be assayed may be one which, in its original state, absorbs only weakly and perhaps only in the far UV where many other species absorb. It would then be necessary to carry out a chemical interconversion, to produce a species which absorbs quite strongly in a distinctive wavelength range. Ideally, this absorption band should be specific to one particular chemical species.

As an example, we may cite the estimation of iron(II) in the presence of iron(III) ion. The former ion, in its aquated form, shows very little absorption even in the UV. By adding a stoichiometric excess of the reagent 1,10-phenanthroline, at a controlled pH, iron(II) is converted into the complex, tris(1,10-phenanthroline) iron(II), which absorbs strongly in the visible with a peak at 510 nm, where $\varepsilon = 1.10 \times 10^4$ dm^3 mol^{-1} cm^{-1}. The intensity of this absorption is attributable to the interaction of the π-electron system of the ligand with the d orbitals of the metal ion.

In using this method, it is desirable to construct a calibration curve using several different known concentrations of iron(II). All aliquots should be treated in the same way and the absorbance at 510 nm is plotted against the concentration of iron(II), giving a straight line, virtually through the origin. The solutions with the unknown concentrations should then be treated in the same way, and their concentration deduced from the calibration curve. In addition, it would be desirable to run the spectrum of each prepared sample to check that, over the relevant wavelength range, the only absorption band present is that peaking at 510 nm.

4.7 Nuclear magnetic resonance spectroscopy

Most, but not all, nuclei have a finite nuclear spin, I. When such a nucleus is subjected to an external magnetic field, it will align itself with respect to the field in one of $(2I+1)$ different ways. Where $I = \frac{1}{2}$, as it is for the hydrogen atom, ^1H, there are then two possible values for the magnetic quantum number, m, namely $+\frac{1}{2}$ and $-\frac{1}{2}$. The difference in the energies of these two states is proportional to the strength, B_o, of the magnetic induction field, as is illustrated in Figure 4.7.

With the magnetic field applied, the population of the state with $m = +\frac{1}{2}$ is very slightly greater than of that with $m = -\frac{1}{2}$. To achieve a transition between these states, a quantum of the correct size is needed. With magnetic fields of a few tesla, the frequency of the required quantum is of the order of several tens to a few hundreds of MHz: that is, the appropriate radiation belongs to the radio range.

As we shall see, an NMR spectrometer needs not only an intense magnetic induction field, but one of high uniformity. To diminish the effects of any variations in the magnetic induction field at the location of the sample, a widely

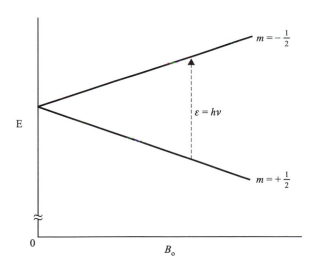

Figure 4.7 Illustration of the effect of an external magnetic field on the two possible energy states of the nucleus of a hydrogen atom, ^1H.

used technique is to rotate the sample at its position between the poles of the magnet. In this way, each molecule of the sample will experience, over a short period of time, an averaged magnetic environment.

In 1951 it was shown that when an organic molecule with quite distinct groups of H atoms was used as a sample, the conditions of resonance were satisfied at perceptibly different values of the applied magnetic field. Thus, ethanol produced three different peaks, for the protons of the methyl group, of the methylene group and of the hydroxyl group. This phenomenon is known as the chemical shift and its discovery clearly indicated that NMR spectroscopy would become a powerful tool in the determination of molecular structure.

The origin of the chemical shift is that the behaviour of the nucleus depends on the local magnetic induction field experienced at that nucleus. This slightly differs from B_o because of the shielding effected by the surrounding electron cloud. Since the extent of the electron density and thus of the shielding varies with the chemical environment of the H atom, the resonance conditions for the different groups of protons will be satisfied at different values of B_o, depending on their chemical environments.

One extremely useful facet of NMR spectroscopy is summed up in the fact that the areas of the three peaks detected for ethanol are in the ratio of 3:2:1. This reflects the fact that NMR absorptions are equally intense for all H atoms, so that the magnitude of the peak is a reliable indicator of the relative numbers of protons involved. Consequently, most NMR instruments have an integration facility for measuring peak areas, since a knowledge of this is an essential aid to determining the molecular structure.

For the most part, NMR spectra are run on solutions where the compound of interest is the solute. An appropriate solvent is then one like CCl_4 or $CDCl_3$, with no H atoms present, provided that the compound is sufficiently soluble in it.

4.8 NMR and structure determination

To facilitate the quantitative use of the chemical shift, a numerical scale was established of such a nature that it is equally applicable to spectra run on any instrument, whether 60 or 400 MHz. The first step was to establish a reference point and for this purpose the signal from tetramethylsilane, $Si(CH_3)_4$ (TMS) was adopted. This compound is chemically inert, gives a single sharp peak at a higher value of B_o than that from nearly any H atom in an organic molecule, and it can be added to the usual solution sample to act as an internal standard.

The scale itself is dimensionless and may be defined as,

$$\delta = \left(\frac{B_{TMS} - B_s}{B_{TMS}} \equiv \right) 10^6 \qquad (4.7)$$

where B_{TMS} and B_S are the values of the magnetic induction field at which the peak of TMS and a particular peak of the sample come into resonance.

Consequently, the TMS peak comes at $\delta = 0.0$. For nearly all H atoms attached to an organic molecule, B_S is less than B_{TMS} so that the δ value is positive.

An NMR spectrum is normally run with the magnetic induction field increasing from left to right along the abscissa. The characteristic ranges of the chemical shift for various types of hydrogen atom are listed in Table 4.2. In using these data, it is important to realise that the δ value for any particular type of H atom can be distorted by its chemical environment. In ethane, the resonance of a methyl hydrogen occurs at $\delta = 0.9$, but in fluoromethane it is at $\delta = 4.3$. The reason is that the F atom, with its high electron affinity, pulls so much electron density towards itself that this deshields the nuclei of the H atoms to a considerable extent, so that the value of the field B_o which needs to be applied to achieve the requisite field at the H atom is much less for CH_3F than for C_2H_6. The ranges quoted in Table 4.2 assume a hydrocarbon environment.

An example of an NMR spectrum is shown in Figure 4.8. This shows three main groups of protons, with the middle group neatly subdividing into two, both of which are clearly aromatic. In terms of the values, the peaks are centred at 9.9, 7.6, 7.25 and 2.4. The first of these is in the aldehydic range while the last is clearly alkyl. The integrated intensities of the peaks fall in the ratio 1:2:2:3.

Clearly, to satisfy the structural evidence we require a molecule with a disubstituted benzene ring, an aldehydic group and a methyl group. Furthermore, the aromatic protons are divided in the manner that is characteristic of an aromatic ring that is substituted in the *para*-positions. In fact, the compound is *p*-methylbenzaldehyde.

Table 4.2 Approximate values of the chemical shift for hydrogen atoms in specific locations

	δ
$C-CH_3$	0.9
$C-CH_2-C$	1.5
$R-NH_2$	1.5
$R-CO-CH_3$	2.1
$R-C\equiv CH$	2.2
$O-CH_3$	3.3
$R-CO_2-CH_3$	3.7
$R-OH$	3.8
$C=CH_2$	5.0
$Ph-OH$	6.0
$Ph-H$	7.3
$R-CO-NH_2$	5.5–6.0
$R-CHO$	9.2–11.0
$R-CO_2H$	9.8–11.8

R denotes an alkyl group such as ethyl, C_2H_5, and Ph the phenyl group, C_6H_5.

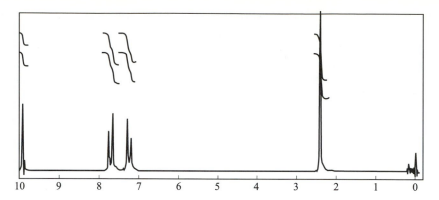

Figure 4.8   An NMR spectrum of *p*-methylbenzaldehyde, $CH_3.C_6H_4.CHO$, run on a 90 MHz instrument, with a drop of tetramethylsilane added as an internal standard. The integration of the signals shows that their areas are in the ratio of 1:2:2:3.

4.9   Fine structure in NMR spectra

When the technique of NMR had advanced a little beyond the stage referred to earlier, it was realised that the main peaks in the NMR spectra of simple molecules like ethanol are actually split into component peaks. This is interpreted as a consequence of the interactions between neighbouring magnetic nuclei via the bonding electrons and usually termed spin–spin splitting.

A nucleus with an adjacent H atom will experience one of two possible magnetic fields depending on whether, for that H atom, $m = +\frac{1}{2}$ or $m = -\frac{1}{2}$. In consequence, the original nucleus will resonate at one of two slightly different applied fields. The resulting absorptions will be of equal size because, statistically, half the H atoms will have $m = +\frac{1}{2}$ and half $m = -\frac{1}{2}$.

If there are n neighbouring H atoms that are chemically and magnetically equivalent, the number of resonance lines will be $(n + 1)$. The weightings of these lines will be as follows:

$n = 1$	weighting $=$	1:1
2		1:2:1
3		1:3:3:1
4		1:4:6:4:1
5		1:5:10:10:5:1

These sets of numbers are known as the binomial coefficients, since they appear in the expansion of $(a + b)^n$, for $n = 1, 2, 3$ etc.

Thus in ethanol, the peak arising from the methyl group is split into three by the two H atoms of the methylene group. Likewise, the absorption from the methylene group is split into four by the three H atoms of the methyl group. In addition, each of these four peaks is split, by a much smaller amount, into two by the H atom of the hydroxyl group.

Regarding the spin–spin splitting by neighbouring nuclei, one limitation is that a resonance does not show fine structure from the interaction of magnetically equivalent nuclei. This explains why the twelve protons of tetramethylsilane (TMS) give only a single peak, with no fine structure. In regard to its role as a standard, this fact is obviously advantageous.

Splittings can also be caused by various other magnetic nuclei, such as $^{13}C(I = \frac{1}{2})$, $^{14}N(I = 1)$ and $^{19}F(I = \frac{1}{2})$. The nuclides ^{12}C and ^{16}O, the major naturally occurring isotopes of these elements, both have $I = 0$, since they have even values of Z and of N. While spin–spin splitting clearly conveys additional structural information and thereby assists a structure determination, if the most abundant isotope of carbon were to have been endowed with a finite nuclear spin, the NMR spectra of organic compounds would be much more complex than in fact they are.

Suggested reading

SANDERS, J. K. M. and HUNTER, B. K., 1987, *Modern NMR Spectroscopy: A Guide for Chemists*, London: Oxford University Press.

BANWELL, C.N. and McCASH, E., 1994, *Fundamentals of Molecular Spectroscopy*, 4th Edn, London: McGraw-Hill.

HOLLAS, J. M., 1991, *Modern Spectroscopy*, Chichester: Wiley.

WILLIAMS, D. H. and FLEMING, I., 1980, *Spectroscopic Methods in Organic Chemistry*, 3rd Edn, London: McGraw-Hill.

SILVERSTEIN, R. M., MORRILL, T.C. and BASSLER, C., 1991, *Spectrometric Identification of Organic Compounds*, 5th Edn, New York: Wiley.

HARRIS, R. K., 1986, *Nuclear Magnetic Resonance Spectroscopy*, London: Longman.

ABRAHAM, R. J., FISHER, J. and LOFTUS, T. P., 1988, *Introduction to NMR Spectroscopy*, Chichester: Wiley.

Problems

4.1 Express in electron volts the energies of quanta of radiation of (a) 3200 cm^{-1}, (b) 550 nm and (c) 90 MHz.

4.2 For the HF molecule, the force constant, k, has been measured as 960 N m^{-1}. Evaluate, in cm^{-1}, the position of the vibrational absorption band. Also, evaluate the zero-point energy of the HF molecule, in electron volts.

4.3 Liquid water, H_2O, has an extremely weak absorption band at 760 nm, which is one reason why the sea can be seen to be blue. Given that D_2O shows no comparable band at wavelengths below 900 nm, deduce whether this arises from an electronic or from a vibrational transition.

4.4 An aqueous solution of KMnO$_4$, of concentration 1×10^{-4} mol dm^{-3}, shows an absorption peak at 525 nm, where only 14.7% of the incident light is transmitted through a cell of 4-cm path length. At this wavelength, evaluate the absorbance, A, and the molar decadic absorption coefficient, ε.

4.5   An unknown substance, boiling at 178°C, has a proton NMR spectrum showing two groups of peaks at 7.3δ and at 9.2δ , whose integrated areas are in the ratio of 5:1. The IR spectrum shows an intense band at 1700 cm^{-1}, and the molar mass has been estimated as 105 ± 10 g mol^{-1}. Suggest a possible structure.

4.6   To ascertain the concentration of hydrogen peroxide in a solution, an aliquot of 10 cm^3 was added to an acidified solution of 10^{-1} mol dm^{-3} KI, where the reaction,

$$H_2O_2 + 3I^- + 2H^+ = I_3^- + 2H_2O$$

would occur quantitatively. This solution was made up to 50 cm^3 with distilled water. It was then used to fill a spectrophotometer cell of 1-cm path length, and the UV spectrum was found to have a peak at 350 nm, where the absorbance, A, was 0.92. Given that for I_3^- ion, $\varepsilon_{max} = 2.6 \times 10^4$ dm^3 mol^{-1} cm^{-1} at 350 nm, evaluate the concentration of H_2O_2 in the original solution.

5

States of matter and intermolecular forces

Here we consider the different states, namely gaseous, liquid and solid, in which chemical substances may exist, the salient properties of each phase and the conditions for the coexistence of two or more phases. For a substance consisting of discrete molecules, it is shown that the deviation of a gas from the perfect gas equation is a direct consequence of intermolecular attractions, either of the Van der Waals type or of a more specific type of interaction called hydrogen bonding. These forces are also in evidence in regard to the behaviour of liquids. The important role of hydrogen bonding in solids and, intramolecularly, in certain macromolecules is also highlighted.

5.1 The gaseous state

A gas totally fills any volume in which it is contained and it exerts a uniform pressure on all the walls of the container. On the laboratory scale, gravity has no effect on the disposition of a gas within the available volume. However, on the macroscopic scale, this is not so: the atmosphere is appreciably 'thinner' at the top of Mt Everest than it is at sea level.

Studies of the behaviour of gases, even in a purely physical way, proved to be very important in the development of chemical theory. However, at many stages there were difficulties to be overcome, in solving the technical problems involved in handling gases. A discussion of these is beyond the scope of this book.

In 1661, Boyle showed that, when the temperature is kept constant, the pressure, P, of a fixed amount of gas is inversely proportional to the volume, V, the gas occupies. This finding can be rationalised on the basis that the gas molecules are perpetually in motion and are continually undergoing elastic collisions with each other and with the walls of the vessel. Thus when the accessible volume is halved, there will be twice as many molecules per unit volume and consequently twice as many collisions per unit surface area on the walls of the vessel. Since the gas pressure is the direct result of the change in momentum effected by these collisions, halving the volume should cause the pressure to be doubled.

The foregoing assumes the presence of a single pure gas. Where a mixture of gases is present, then each must make its own contribution to the perceived pressure and the total pressure is the sum of a set of partial pressures, one in respect of each component of the mixture. Each partial pressure may be regarded as the pressure that that gas would exert if it alone were present in the same total volume.

This concept of partial pressures is totally in harmony with the qualitative model referred to above. Each partial pressure may be regarded as the consequence of the collisions with the walls made by the molecules of that particular gaseous component. A pressure measuring device, such as a manometer, will measure only the total pressure, so that other means must be employed to determine the individual partial pressures.

Later studies of the effect of temperature showed that, at constant pressure, the volume occupied by a fixed amount of gas is proportional to the temperature, T, on the Kelvin scale. (In practice, this is a figure greater by 273.15 than the Celsius temperature: water freezes at 273.15 K.) When this finding, whether called Charles' Law or the Law of Gay-Lussac, is combined with Boyle's Law we have the result that PV/T is a constant for a fixed amount of gas. It was proposed by Avogadro, and eventually accepted as true, that equal volumes of all gases, at the same temperature and pressure, contain equal numbers of molecules. Since the quotient, PV/T is proportional to the amount of gas involved, with the assistance of Avogadro's hypothesis we may obtain the equation:

$$PV = nRT \qquad\qquad (5.1)$$

This is known as the equation of state for a perfect gas, where n is the number of moles of gas present and R is called the gas constant. This fundamental constant, equal to PV/nT, has the value, in SI units, of 8.3145 J K^{-1} mol^{-1}.

In using Equation (5.1), it is necessary that all quantities are in consistent units. The SI unit of pressure, called the pascal and equal to one newton per metre squared, is rather small by terrestrial standards. The pressure of a standard atmosphere, defined as that supporting a column of mercury 760 mm high, is 1.01325×10^5 N m^{-2}. So the volume occupied by one mole of gas at 298 K and one atmosphere pressure is given by:

$$V = \frac{RT}{P}$$

$$= \frac{8.3145 \text{ J K}^{-1}\text{mol}^{-1} \times 298 \text{ K}}{1.01325 \times 10^5 \text{ N m}^{-2}}$$

$$= 0.02445 \text{ m}^3 \text{ mol}^{-1}$$

$$= 24.45 \text{ dm}^3 \text{ mol}^{-1}$$

(5.2)

This answer shows how the density of a substance varies in different phases. One mole of a liquid or a solid might occupy 20–40 cm^3. At atmospheric pressure, a gas occupies approximately a thousand times this volume, so that the actual volume of the gas molecules themselves must be quite small in relation to V.

5.2 The kinetic theory of gases

The model of a gas referred to in the previous section deserves to be amplified in some respects. The gas molecules are perpetually in motion, undergoing collisions with each other and with the walls of the vessel. It may be deduced that, at any moment in time, the molecules do not all possess the same velocity. Rather, as is shown in Figure 5.1, there is a distribution of velocities, from zero upwards.

The distribution function for the velocity, c, depends on two parameters, the molar mass, M, and the Kelvin temperature, T. The maximum of this curve corresponds to the most probable velocity, c_{mp}, which occurs at the value

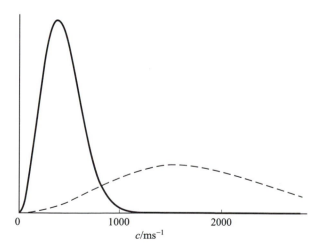

Figure 5.1   The distribution of velocities of O$_2$ (full line) and of H$_2$ (dashed line) at a temperature of 298 K.

$(2RT/M)^{1/2}$. Since the curve is skewed to the right, the mean velocity, c, is a little greater than this and is given by $(8RT/\pi M)^{1/2}$.

It is of interest to substitute numbers into this expression to see the order of magnitude of molecular velocities. Suppose we have gaseous N_2 at 298 K. The molar mass is 28 g mol^{-1} = 0.028 kg mol^{-1}, and the mean velocity is given by:

$$\bar{c} = \left(\frac{8RT}{\pi M}\right)^{1/2}$$

$$= \left(\frac{8 \times 8.314 \text{ J K}^{-1}\text{mol}^{-1} \times 298 \text{ K}}{\pi \times 0.028 \text{ kg mol}^{-1}}\right)^{1/2} \tag{5.3}$$

$$= 475 \text{ m s}^{-1}$$

This answer, which equates to 1062 miles per hour, is of the same order of magnitude as the speed of sound.

The kinetic energy of a molecule of mass m is given by $\frac{1}{2}mc^2$. So to evaluate the mean kinetic energy of the molecules, we need the mean value of c^2, which is given by $3RT/M$. Thus the square root of the mean value of c^2, called the root mean square, $\sqrt{c^2}$, is given by $(3RT/M)^{1/2}$. Also, the mean kinetic energy of a molecule is given by $3RT/2N_A$, where N_A is Avogadro's constant and is independent of the molar mass. This means that the Kelvin temperature is directly proportional to the kinetic energy of the molecules in the gas phase, and this fact is sufficient justification for this temperature scale.

If the temperature is held constant, the maximum in the velocity distribution curve varies with $M^{-1/2}$. This is illustrated in Figure 5.1, where the full line refers to O_2 molecules at 298 K, and the dashed line to H_2 molecules at the same temperature. For the latter, the values of c_{mp}, $\bar{c}$ and $\sqrt{c^2}$ are all greater than for O_2 by a factor of 4.

In the Earth's upper atmosphere, where the air is thinner, it is conceivable that a molecule may acquire a sufficiently high velocity to escape the Earth's gravitational field. As Figure 5.1 illustrates, the lighter a molecule is, the more likely it is to possess a very high value of c in the 'escape velocity' range. For this very reason, the Earth's atmosphere contains virtually no hydrogen or helium. While the escape of N_2 or of O_2 is not totally precluded, the probability of either is so much less that there is negligible depletion of these components of the Earth's atmosphere.

In making quantitative estimates of any of the parameters related to molecular motion in gases, a useful concept is that of the mean free path, λ, which denotes the mean distance that a molecule travels between collisions. This depends, obviously, on the size of the molecules and on the gas pressure, in an inverse manner in both cases.

$$\lambda = \frac{RT}{\sqrt{2}\pi d^2 N_A P} \tag{5.4}$$

If we substitute $d = 0.37$ nm $= 3.7 \times 10^{-10}$ m, which is the value for N_2, and $P = 1.013 \times 10^5$ N m^{-2}, which is atmospheric pressure, this yields at 298 K a value of $= 6.7 \times 10^{-8}$ m $= 67$ nm. While this may seem a very short distance, it is in fact equal to 180 molecular diameters. This illustrates the fact that, at the molecular density corresponding to one atmosphere pressure, molecules move a distance far in excess of their linear dimensions between successive collisions.

5.3 Deviations from the equation of state for gases

While Equation (5.1) correctly expresses the combined findings of the scientific pioneers and is followed, within experimental error, by most permanent gases at or above room temperature and at pressures less than atmospheric, it is not universally followed by all gases and vapours under all experimental conditions. In particular, large deviations are found at very high pressures and low temperatures, under conditions which were scarcely accessible in the early studies.

There are two main reasons why the behaviour of gases should deviate from that detailed in Equation (5.1). Firstly, in our qualitative model of a gas, the volume of the actual molecules is assumed to be negligible in relation to V, the volume to which the gas has access. While Equation (5.2) shows that this ratio is indeed small at atmospheric pressure, at much higher pressures V will have fallen to the extent that the actual volume is no longer a negligible fraction.

Hitherto, we have treated our gas molecules as if they were like golf balls, exerting no forces on each other except when they actually touch, in which case they repel each other. Some molecules, such as two helium atoms, behave in a manner very close to this, but in general there are also forces of attraction between a pair of molecules, perceptible even when they are not touching. Both of these factors were recognized by Van der Waals in his classic paper in 1873 dealing with the continuity of the liquid and gaseous states.

One means of describing the deviations of real gases from the 'ideal' behaviour represented by Equation (5.1) is to use the virial equation of Kamerlingh Onnes. For one mole of gas, this is,

$$\frac{PV}{RT} = 1 + \frac{B}{V} + \frac{C}{V^2} + \cdots \tag{5.5}$$

where B is called the second virial coefficient, C the third virial coefficient and both are dependent on temperature. (The word 'virial' is derived from the Latin, *vis* = force.)

To evaluate B, let us rearrange Equation (5.5), relating to one mole of gas:

$$\left(\frac{PV}{RT} - 1\right)V = B + \frac{C}{V} + \cdots \tag{5.5a}$$

For experiments where the pressure is varied at constant temperature, if the left-hand side is plotted against V^{-1}, then B is obtained as the intercept and (to an approximation) C as the slope.

Clearly, for a perfect gas B and C will both be zero. A sophisticated theoretical treatment of gases, using the methods of statistical thermodynamics, shows that the second virial coefficient, B, may be correlated with the attractive interactions between pairs of molecules at separations slightly greater than the molecular diameter. If there were no attractive forces operative between pairs of molecules, then B would be expected to be zero, as also would C.

5.4 Intermolecular (Van der Waals) forces

We will discuss these in terms of the interactions of pairs of molecules, which is implicitly claiming that the total effect is well approximated by the sum of the interactions between individual pairs. Secondly, to specify how pairs of molecules interact, we will use plots of the potential energy, U, as a function of the separation r, of the centres of the two molecules, taking $U = 0$ at infinite separation. If the molecules repel, then there is energy stored in the pair of molecules and U is positive; where they attract, U will be negative.

If the two molecules behave as hard spheres, then as they approach each other, U remains at zero until they touch at $r = R$, when U rises vertically, as shown in Figure 5.2(a). A more realistic type of repulsion has U rising more gradually, and with increasing steepness, as in Figure 5.2(b). If two helium atoms were brought together so that the electron clouds start to overlap, the forces of repulsion would be expected to rise in a manner of this type. In fact, for any two molecules that do not undergo a chemical reaction when they collide, one would expect, at very short separations, forces of mutual repulsion consistent with the plot of U shown in Figure 5.2(b).

In general, when two molecules are at a rather greater separation, U will be negative because of forces of mutual attraction. When these are taken account of then the overall potential varies with separation as shown in Figure 5.2(c).

In a single substance, there are two alternative causes of intermolecular attraction. If there is a plane about which the molecule is not symmetrical, so that it has a permanent dipole (as have HCl, SO_2 or CH_3Cl), then these dipoles will interact in a manner which depends on orientation. However, if the molecules are free to rotate, then the attractive orientations will predominate and U is negative and proportional to r^{-6}.

Some molecules, such as Ar, N_2, CO_2, CF_4 or C_6H_6, have sufficient symmetry that they possess no permanent dipoles. In a mixture of Ar with a molecule of the previous group, the approach of the permanent dipole would induce a weak dipole in the symmetrical molecule, leading to mutual attraction. However, in a pure substance, the dipole-induced dipole interaction is unimportant.

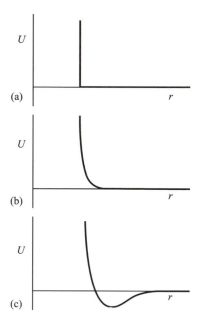

Figure 5.2   Plots of the potential energy, U, as a function of the separation, r, for the cases of: (a) a hard sphere potential, (b) a repulsive potential, attained more gradually and (c) a more characteristic potential, with attractive and repulsive forces involved.

While an atom like argon or a molecule like dinitrogen possesses a high degree of symmetry, in each the electrons are continually in motion so that there will be instantaneous dipoles. The interaction of two instantaneous dipoles leads to attractive forces between the molecules, usually known by the name of **dispersion** forces. These also give a negative value for U, proportional to r^{-6}, but the magnitude is much less than that between two permanent dipoles.

Both the dipole–dipole and the dispersion interactions between molecules are spoken of as short range forces. In each case, when r is doubled, U decreases to 1.6% of its previous value. This contrasts with the Coulombic forces between two ions, where U is proportional to r^{-1} and so it falls off only very slowly with increasing separation. These are thus long range forces.

There is one further type of force which may act between molecules, in addition to the categories considered above. Hydrogen bonding is a much more specific type of interaction, exhibited only by certain classes of molecule, so it is convenient to leave it to be described in a later section. However, it may be argued that hydrogen bonding is even more significant than dipole–dipole or dispersion forces.

5.5 The liquid state

A liquid, like a gas, is a fluid, but it may not fill the vessel which encloses it. Another difference is that a liquid readily demonstrates the effects of gravity. If a pure substance is put inside a container, then one may have a liquid phase at the bottom, with a horizontal surface separating it from the gas or vapour phase above it.

A liquid, unlike a gas, is a condensed phase, and around most molecules there will be a number of others at a separation close to that at which U has its minimum value. Thus, work has to be done to separate the molecules of a liquid, and this is the origin of the energy, called the latent heat of vaporisation, required to convert a liquid into a vapour at the same temperature. However, two points are important. The first is that the make-up of a liquid is not static, so that no molecule keeps its nearest neighbours. There is continual movement and rearrangement. The second, which follows naturally, is that while there is some order at any point in time immediately around a molecule, there is no long-range order, even on a very brief time scale.

For a pure substance, the pressure of the vapour in equilibrium with the liquid rises as the temperature is increased. In comparing the behaviour of different substances, there are two alternative bases: one may compare the vapour pressures at a common temperature or, alternatively, one may compare the boiling points, which are the temperatures at which the vapour pressures attain one atmosphere. In terms of availability of data, the latter is preferable and a low boiling point means high volatility.

Boiling points usually demonstrate two trends. One is that, for molecules involving bonds of a particular type, the boiling point increases as the molar mass increases. The rationale is that the greater the molar mass, the smaller, at constant temperature, is the mean velocity. Consequently, the heavier the molecule, the more slowly it moves and the more readily it will condense: in other words, the higher will be the boiling point. This is readily exemplified by the alkanes, of formula C_nH_{2n+2}, where the boiling points of methane (CH_4), ethane (C_2H_6), propane (C_3H_8), n-butane (C_4H_{10}), and n-pentane (C_5H_{12}) are $-164°$, $-89°$, $-42°$, $-0.5°$, and $+36°C$ respectively, and show a continual rise. The other trend is that the boiling point rises with increasing polarity of the molecule. Thus the non-polar 1,4-dimethylbenzene boils at $138°C$, 1,3-dimethylbenzene (which is slightly polar) at $139°C$ and 1,2-dimethylbenzene (which is rather more polar) at $144°C$.

5.6 Viscosity and self-diffusion in liquids

Liquids are characteristically capable of flowing, but in this respect they exhibit a huge range of behaviour, with some flowing easily and others only ponderously slowly. The relevant property is called the viscosity, which is the frictional resistance to an applied shearing force. The layer of a liquid in contact with a surface is stationary, so that if the liquid above this is flowing, there must be a velocity gradient. The frictional force is proportional to this gradient and to the area of the adjacent layers, and the relevant constant of proportionality is called the coefficient of viscosity, η, whose SI unit is 1 kg m^{-1} s^{-1}.

The viscosity coefficient can be measured either by the rate of flow of the liquid through a capillary or by the force required to rotate a solid cylinder inside a concentric cylinder filled with the liquid. The viscosity coefficient of methanol at 298 K has the value $\eta = 0.547 \times 10^{-3}$ kg m^{-1} s^{-1}, whereas those of ethanol (1.09×10^{-3}) and n-butanol (2.6×10^{-3}) show a steadily increasing trend. In general, η decreases with increasing temperature, in such a manner that ln η is a linear function of T^{-1}

For small non-polar molecules, with only dispersion forces of attraction between them, a low viscosity coefficient is to be expected, since there is no particular obstacle to dragging one molecule past another. Thus for n-hexane, η has the low value of 0.29×10^{-3} kg m^{-1} s^{-1}. In alcohols, where there is strong intermolecular hydrogen bonding, the η values of the lower members are quite modest, but clearly, a C_n alcohol has a much higher viscosity coefficient than the corresponding C_n alkane. A much larger effect is to be found where the number of hydroxy groups is increased with the size of the molecule. This is illustrated in the series, methanol (CH_3OH), ethanediol ($HOCH_2CH_2OH$) and propanetriol ($HOCH_2CH_2(OH)CH_2OH$) where η (in the units of 10^{-3} kg m^{-1} s^{-1}) rises from 0.547 to 16.2 to 954.

A related property of fluids is the random independent motion of the molecules. If a solute is present, located non-uniformly in the solution, then molecular motion will work towards achieving uniformity. Across any plane, the net rate of motion is proportional to the concentration gradient. The constant of proportionality is the diffusion coefficient, D, with units m^2 s^{-1}. The diffusion of a solute reflects that of the solvent and, where the two molecules are similar in size, their respective diffusion coefficients are very similar. The diffusion coefficients of large solute molecules are much smaller.

For the diffusion of water in water, the relevant diffusion coefficient, called the coefficient of self-diffusion, is 2.1×10^{-9} m^2 s^{-1} at 298 K. The diffusion coefficients of many other species in water are around this value. If a drop of a concentrated aqueous solution of such a species were to be deposited without any agitation in the middle of a beaker of water, the root-mean-square distance, l, that the molecules of solute would diffuse in time t is given by:

$$l = (6Dt)^{1/2} \tag{5.6}$$

If D is the same as the coefficient of self-diffusion of water, we may deduce that for a period of 1 hour, l is merely 6.8 mm. Clearly, achieving a homogeneous solution will be a very slow process unless there is substantial assistance from convection currents or by stirring.

A theoretical analysis by Stokes interrelated the coefficients of diffusion and of viscosity, D and η, but in an inverse manner. DT/η should be a constant. Usually, diffusion coefficients increase with increasing temperature, with ln D

being an essentially linear function of T^{-1}. However, it is much easier to make an accurate measurement of the coefficient of viscosity than of the diffusion coefficient.

5.7 The solid state

The other condensed state of matter is the solid state. The ideal solid state is a perfect crystal, in which the atoms/ions/molecules that compose the crystal occupy recurring ordered positions in a three-dimensional array. It is thus highly ordered, with this order extending to long ranges. Unless crystals have been carefully grown, the solid may very well be polycrystalline, or made up of separate ordered regions, each with different three-dimensional axes. These ordered regions, or crystallites, are joined together at grain boundaries.

Various different types of crystalline solids can be distinguished, on the basis of the nature of the forces holding the particles together. However, this subdivision is not absolute and some solids do not seem entirely to fit any single class.

5.7.1 Covalent solids

The most familiar example of a solid in which the atoms are bonded together by covalent forces is diamond, an allotrope of carbon. Here, the atomic orbitals of each carbon atom are sp^3 hybridised, and σ bonds are formed between carbon atoms, just like the C–C bonds in ethane or propane. This gives a rigid three-dimensional structure, as illustrated in Figure 5.3, and it is for this reason that diamond is one of the hardest known substances. It also has a very high melting point.

5.7.2 Ionic solids

Examples of these are given in Figure 3.1, which depicts the crystal structures of NaCl and CsCl. In ionic solids, ions occupy positions in a lattice so that the nearest neighbours of any ion are of the opposite sign, and are attracted to it. The next nearest neighbours are ions of the same sign, which will repel; however, the attractive interactions are predominant. Ionic solids usually have fairly high melting points.

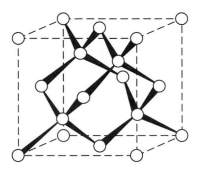

Figure 5.3 An illustration of the structure of diamond, in which each carbon atom is bonded to four other C atoms arranged tetrahedrally.

5.7.3 *Molecular solids*

Let us consider a 'typical' organic molecule, in which various atoms of C, H, O and perhaps other elements are covalently bonded together to give a discrete molecule. When this compound forms a crystalline solid, with a three-dimensional array of molecules in a regular repeating pattern, then they are held together only by the Van der Waals forces described in section 5.4. Consequently, such compounds usually have low melting points. For example, benzene (C_6H_6), melts at 5.5°C, perchloropropane (C_3Cl_8) at 160°C and carbon disulfide, CS_2, at −111°C.

The melting points of this group of substances tend to show the same general effects as the boiling points. However, the correlation between polarity and melting point is not necessarily a good one. For the series 1,4-, 1,3- and 1,2-dimethylbenzene, the melting points are 13, −48 and −25°C respectively. The obvious reason is that whereas volatility is a reflection only of intermolecular forces, packing considerations are also important in regard to the equilibrium between the liquid and the solid states.

5.7.4 *Metals*

Metals differ from other elements, particularly in respect of their high electrical conductivity. They have a unique form of bonding, which is associated with the fact that these elements have fairly low ionisation potentials. To an approximation, they may be described by assuming that the ions formed by the loss of those electrons which are easily removed are located at certain fixed sites and then surrounded by a 'sea' of electrons, which serves both to bind the ions together and to provide the high electrical conductivity.

Several different crystal structures are exhibited by metals. Many adopt one of the close packed arrangements, in which each atom has twelve nearest neighbours. Examples include metals like platinum and lead. A few adopt the body centred cubic structure with eight nearest neighbours, and this group includes iron and tungsten. The melting points of metals vary widely.

5.8   Phase diagrams and the coexistence of phases

Most substances are capable of existing in the solid, the liquid and the gas or vapour phase. The phase or phases actually present depends on the conditions of temperature (T) and pressure (P). If an amount of a pure substance is put inside a closed vessel in which these parameters may be controlled, then the phases present at any combination of P and T may be summarised in a phase diagram, of which a hypothetical example is shown in Figure 5.4.

This figure shows, for example, that at the conditions prevailing at the letter G, only the vapour will be present. At the letter L, there is only liquid and at letter S, only solid. The lines denote the various conditions for the coexistence of two phases. The line between liquid and vapour has a positive slope, reflecting the fact that the vapour pressure of a liquid increases with increasing temperature. The line denoting the coexistence of solid and vapour also has a positive slope, for a similar reason. However, there is a change of slope where these two lines join for the reason, which is more fully explained in Chapter 7,

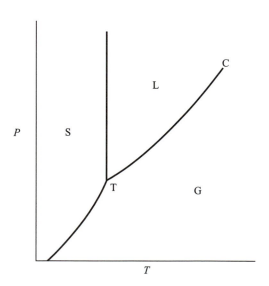

Figure 5.4   A hypothetical phase diagram of a single substance, on a plot of the pressure, *P*, against the temperature, *T*. The point C denotes the critical point and T the triple point.

that the latent heat of sublimation of a substance is greater than its latent heat of vaporisation.

The line representing the coexistence of solid and liquid is usually nearly vertical, with its slope depending on the relative densities of these two condensed phases. The point, T, where this line intersects the other two is called the triple point and specifies the unique conditions necessary for the coexistence of all three phases. The triple point of water occurs at a pressure of 4.58 mmHg (or 4.58 torr) and a temperature of 298.16 K, and this latter figure is used to define the Kelvin scale of temperature. The value of the triple point pressure is related to the behaviour of the substance at atmospheric pressure. The above value for water is much less than one atmosphere; consequently, water has a large liquid range. For some other substances, the triple point pressure exceeds one atmosphere, consequently substances like CO_2 or I_2 have no liquid range at atmospheric pressure, and on heating they simply go from the solid to the gas or vapour phase.

In Figure 5.4, the lines between liquid and solid and between vapour and solid both carry on indefinitely. On the other hand, the line between liquid and vapour only goes so far. At temperatures higher than this, it is not possible to have the coexistence of vapour and liquid phases, no matter how great a pressure is applied. The terminus of this line, denoted by C, is called the critical point. Above that, only the gas phase can exist.

One physical manifestation of the critical point is that each substance has a temperature above which it cannot be liquefied. This was first demonstrated by Andrews in his studies on carbon dioxide in 1869. When the pressure was plotted against the volume at constant temperature, below the critical temperature, T_c, these isotherms had a horizontal portion over which vapour and liquid were present in varying proportions. At higher temperatures, the slope of these plots was consistently negative, reflecting the fact that only one fluid phase was present. At 31°C, the slope attains zero at a point of inflexion, so this is the critical temperature of CO_2. For N_2, for example, the critical temperature is −147°C, or 49°C above the boiling point. Thus air needs to be cooled quite appreciably before the nitrogen can be liquified.

In section 5.3, reference was made to the deviations of real gases and vapours from the equation of state for a perfect gas. If the behaviour of different substances is compared at a constant temperature, then those with a low value of T_c (such as helium, 6.2 K, or hydrogen, 34.2 K) show much less deviation than those with higher critical temperatures. In that sense, the critical temperature of a substance is a parameter of fundamental importance, which reflects the strength of the intermolecular attractive forces.

5.9 Hydrogen bonding

The discussion in section 5.4 of the types of forces between molecules omitted one category, exhibited only by certain types of compound, but arguably of even greater significance than dispersion and dipole–dipole forces. This category is specific to compounds involving atoms of hydrogen, which is in many ways an anomalous element. An H atom bonded to a highly electronegative element may interact quite strongly with another highly electronegative atom and this attraction is known as the **hydrogen bond.**

Let us consider the system,

$$-X-H \cdots Y-$$

where X and Y are both bonded to other atoms so that their normal valencies are satisfied and the dotted line denotes the hydrogen bond between H and Y. If the hydrogen bonding is to be significant, then four points are important:

1 X and Y both need to be elements of high electron affinity, which largely limits them to Groups VII, VI and V.

2 The atoms X and Y should both be small, which means that the strongest hydrogen bonds involve F, O and N from the second row.

3 For effective hydrogen bonding, it is preferred that the atoms $X \cdots H \cdots Y$ are collinear.

4 When hydrogen bonding exists, the separation of H and Y is perceptibly less than one would expect between atoms which cannot interact in this way.

To rationalise these observations, given that X has a high electron affinity, the X–H bond will be strongly polarised, with a net negative charge on X and a net positive charge on H. The atom Y, also highly electronegative, will likewise carry a net negative charge, so that Y can be attracted by the X–H dipole. The energy of this interaction will be the greater the closer H and Y can approach. The restriction of this type of bonding to small X and Y is consistent with the fact that it is peculiar to hydrogen, the element with the smallest atom.

A good illustration of the effect of hydrogen bonding in liquids is provided by the boiling points of the hydrides of Groups IV to VII, shown in Figure 5.5(a). The hydrides of Group IV illustrate the normal behaviour, which is that the boiling point increases with increasing molecular weight. However, for Groups V, VI and VII, the hydrides of the second row elements all have boiling points that are anomalously high. In Figure 5.5(b), the same phenomenon is seen with the molar latent heats of vaporisation of these hydrides.

The most obvious interpretation of Figure 5.5 is that hydrogen bonding is so strong in the compounds NH_3, H_2O and HF that the boiling points and latent heats of vaporisation are raised to a significant extent. While hydrogen bonding effects may also exist in H_2S and HCl, their magnitude is clearly very much less.

Another series which attests to the effect of hydrogen bonding on physical properties is the sequence HOH, CH_3OH and CH_3OCH_3. At each step, the molar mass is being increased by 14, and if this were the only change then

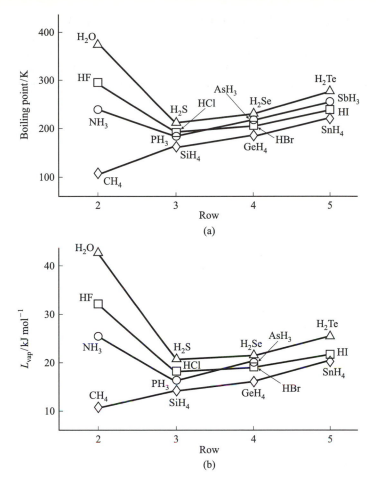

Figure 5.5 Illustration of how (a) the boiling points and (b) the latent heats of vaporisation of the hydrides of Groups IV, V, VI and VII vary with the row number of the element.

one would expect a progressive increase in the boiling point. In fact, the boiling point figures are 100°, 65° and −23°C, respectively, and they show the reverse trend. The explanation is that in advancing along this series, the ability to engage in hydrogen bonding decreases: water has two H atoms which may participate in that way, methanol has one and dimethyl ether has none. Another relevant comparison is of the boiling points of propane (−42°C) and ethanol (78°C). Since the molar masses (44 and 46) are so similar, the considerable difference is attributable to the effect of hydrogen bonding in alcohols.

5.10 Hydrogen bonding in solids and macromolecules

Of all the molecules mentioned above in connection with hydrogen bonding, the one which shows the greatest effects of this phenomenon is H_2O. The reason is well illustrated by reference to the structure of its solid state, ice. This is easily understood on the basis of the MO bonding scheme for water, presented in section 3.5. In this molecule, the O atom is at the centre of a slightly distorted tetrahedron. Two of the sp^3 hybrid orbitals represent lone pairs of non-bonding electrons, with excess negative charge. The other two are engaged in σ bonds with the H atoms, with net positive charge around the H. So each of these H atoms engages in hydrogen bonding with a lone pair on the O atom of another water molecule, just as each lone pair interacts with the H atom of a different water molecule. In this way, each water molecule is hydrogen-bonded to four other water molecules, lying in two different planes, and forming part of a three-dimensional structure, as shown in Figure 5.6.

Two properties of ice follow automatically from the structure described above. Each water molecule is hydrogen-bonded to four others, so the co-ordination number has the very low value of 4 (which contrasts with the value of 12 for a close-packed structure). In consequence, ice has a low density and water is one of only five known substances which expand on freezing. Also, the arrangement of water molecules in ice is almost the same as that of carbon atoms in diamond. In consequence, this structure confers considerable hardness, even though ice is held together only by hydrogen bonds. Taken in concert, these two properties can adequately account for the lethal nature of icebergs: in that sense, the sinking of the *Titanic* may be seen to have been a tragic consequence of hydrogen bonding.

In the solid state, carboxylic acids engage in hydrogen bonding, as illustrated in Figure 5.7. However, the effect is rather different. Each molecule participates in two hydrogen bonds, both to the same molecule.

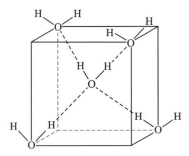

Figure 5.6 An illustration of the crystal structure of ice, where each H_2O molecule is hydrogen bonded to four others, arranged almost tetrahedrally.

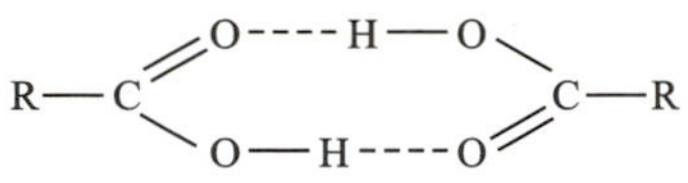

Figure 5.7 A dimer of a carboxylic acid, RCO_2H, formed by two intermolecular hydrogen bonds.

Consequently, a crystalline carboxylic acid is made up of hydrogen-bonded dimers. The dimers are held together by ordinary Van der Waals forces.

Hydrogen bonding plays an important role in many molecules of biological importance. Peptides and proteins are macromolecules, formed from individual amino acids with the recurring unit, –NH–CO–. Peptide chains tend to adopt a coiled rather than a linear configuration, with hydrogen bonding between the $=$N–H of one peptide unit and the $=$C$=$O of another.

The structure of deoxyribonucleic acid (DNA) was elucidated in the early 1950s and this work is one of the few stories of science to be the basis of a best-selling book. The molecule is large and fairly complex but the key to comprehending the structure lay in appreciating the role of hydrogen bonding.

Studies of living cells showed that they contain two kinds of nucleic acids called RNA and DNA. It was suggested that DNA played a key role in the ability of living systems to reproduce themselves. Nucleic acids consist of aggregates called nucleotides, which contain a sugar molecule bonded to a base and these are joined together by phosphate bridges. Chemical studies showed that DNA contained the four bases, adenine (**A**), cytosine (**C**), guanine (**G**) and thymine (**T**). Analysis of DNA from different sources showed that while the proportions of these four were variable, the molar amounts of **A** and **T** were always equal, as were those of **C** and **G**. Whereas **A** and **G** are purine bases, **C** and **T** are pyrimidine bases: this means, of course, that in DNA half the bases are purines and half are pyrimidines.

X-ray crystallographic studies implied that DNA had a helical structure. In 1953, Watson and Crick suggested that the structure was a double helix, with an adenine base unit in one strand hydrogen-bonded to a thymine base unit in the other and likewise cytosine to guanine, as shown in Figure 5.8. The point is that only thymine will fit with adenine and only guanine with cytosine.

It was also proposed that hydrogen bonding controlled the replication of DNA on the basis that this occurred by the separation of the two strands of the double helix and the rebuilding of a second strand for each from its nucleotide components.

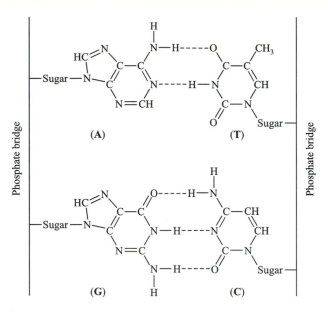

Figure 5.8 A sketch of part of a molecule of DNA (deoxyribonucleic acid) showing how the adenine unit (**A**) forms hydrogen bonds with thymine (**T**), and cytosine (**C**) with guanine (**G**).

Suggested reading

SMART, L. and MOORE, E., 1992, *Solid State Chemistry*, London: Chapman & Hall.

TABOR, D., 1979, *Gases, Liquids and Solids*, 2nd Edn, London: Cambridge University Press.

MAITLAND, G. C., RIGBY, M., SMITH, E. B. and WAKEHAM, W. A., 1987, *Intermolecular Forces: Their Origin and Determination*, London: Oxford University Press.

PIMENTEL, G. C. and McCLELLAN, A. L., 1960, *The Hydrogen Bond*, San Francisco, CA: WH Freeman.

KAUZMANN, W., 1966, *Kinetic Theory of Gases*, New York: Benjamin.

Problems

5.1 An amount of a permanent gas, collected in a Töpler pump, occupies a length of 21.3 mm in a capillary of diameter 1.5 mm, at a pressure of 31.7 torr and at 18°C. How many moles of gas have been collected?

5.2 Evaluate the root-mean-square velocity of a helium atom at 100°C.

5.3 For a gas whose molecules have a diameter of 3.5 Å, evaluate the mean free path at 25°C and a pressure of 10^{-5} torr.

6

Energetics and equilibrium: the Laws of Thermodynamics

Many chemical reactions, such as the combustion of fuels, are accompanied by a release of energy. In consequence, such fuels are spoken of as forms of energy, transportable by ship, truck or pipeline.

A systematic approach to energy changes is put in the context of the First Law of Thermodynamics. This enables thermochemical data to be recorded in such a manner that the energy changes for other reactions, even if hypothetical, may be calculated from them.

A related topic is that of chemical equilibrium. It is shown that thermochemical data alone are insufficient to specify this. The Second Law is presented as a definition of the entropy, S. In spontaneous processes it is shown that S tends to increase and that a function involving S, called the Gibbs energy, acts as a criterion of equilibrium in systems constrained at constant temperature and pressure.

6.1 Work and heat as forms of energy

By Newton's Second Law of Motion, an object which is at rest will remain so unless acted upon by a force. By this Law, the acceleration, a, produced by the application of a force is proportional to the applied force, f, and inversely proportional to the mass, m, of the object. Thus we have,

$$f = ma \tag{6.1}$$

which means that the magnitude of the force may be evaluated from the product of the mass and the acceleration. A force of one newton (N) causes the velocity of a mass of 1 kg to increase by 1 m s^{-1} every second. Thus 1 N may be equated to 1 kg m s^{-2}, and this is the SI (Système International) unit of force.

Mechanical work is done when a force acts through a distance, and the amount of work is given by the product of these parameters. The SI unit, the Joule (J), is the amount done when a force of 1 newton acts through a distance of 1 metre. So $1 \text{ J} = 1 \text{ kg m}^2 \text{ s}^{-2}$.

Alternatively, work may be done electrically. A familiar example is the electric kettle, where a voltage is applied across the heating element and, as a consequence of this electrical work, the temperature of the kettle and its contents rises. This represents an increase in the thermal energy of the kettle plus contents, by an amount equal to the product of the voltage (in volts), the current (in amperes) and the time (in seconds).

Heat is a familiar everyday concept, essentially synonymous with warmth. In science, this word is used with a rather restricted and slightly altered meaning, indicating thermal energy transfer. Thus a kettle on a hotplate attains boiling point by acquiring heat from the hotplate. On the other hand, if we consider the heating element as being an integral part of an electric kettle, the latter achieves the same state by having electrical work done on it.

Quantities of heat were formerly measured in calories, where 1 calorie was the amount of heat required to raise the temperature of 1 g of water by 1 degree. This proved unsatisfactory, chiefly because the amount of heat required for a certain temperature rise depends on the starting temperature. (Put more explicitly, using terms to be met later, because the heat capacity of water is to some extent a function of temperature.) For years, the calorie has been defined in terms of a certain number of Joules. It is now a secondary unity, rarely used in Europe, and with no particular merit apart from its familiarity to those who have never used any other.

Passing heat into a substance usually leads to one of two possible consequences.

1. The substance remains in the same phase and its temperature rises. One usually chooses to do this, *either* at constant volume *or* at constant pressure. The extent of the temperature rise depends on the amount of heat passed in and on a parameter of the substance, called the heat capacity, normally denoted by C, with a subscript V or P to show which of the volume or the pressure is being kept constant. So at constant pressure, we have,

$$q = C_p . \Delta T \tag{6.2}$$

where ΔT denotes the temperature rise. For one mole of a perfect gas, C_p is always greater than C_v by an amount equal to R, the gas constant.

2. The substance changes phase, while the temperature remains constant. The phase change may be solid to liquid, liquid to vapour or solid to vapour, and the heat necessary to achieve this is called the latent heat of fusion, or of vaporisation or of sublimation. These will be denoted by L_{fus}, L_{vap} and L_{sub}.

6.2 The First Law and the enthalpy

We can now proceed to the First Law of Thermodynamics. Suppose we have a certain amount of substance contained within a vessel. Scientifically, this may be termed a closed system. We assume that the system is so disposed that we may do work on it, whether mechanical or electrical, and may pass heat into it. The First Law may then be expressed as,

$$\Delta U = U_f - U_i = q + w \tag{6.3}$$

where q denotes the quantity of heat passed into the system, w the amount of work done on it, U the internal energy of the system, with the subscripts f and i indicating the final and the initial values. This difference is frequently represented by ΔU, where the prefix Δ denotes an appreciable change in the appended quantity.

The internal energy, U, like many quantities used in thermodynamics, is a function of state, dependent only on the state of the system and independent of the route by which the system attains that state. To illustrate this, we may conceive of an all-purpose kettle, fitted both with an appropriate bottom for setting on a hotplate and with a heating element for connecting to an electrical supply. With water present, we may take this system, kettle and contents, from ambient temperature to 100°C either by doing (electrical) work on it or by passing heat into it. So the alternative amounts of work and of heat must be equal.

It is worth noting that, in general, q and w need not be positive quantities. Heat may flow from the system to the surroundings, in which case q is negative. The system might be allowed to expand and so do work on its surroundings: w would then be a negative quantity. Another pertinent point is that, while Equation (6.3) introduces the internal energy, U, it does not define it. Rather, it defines changes in U.

The usefulness of the internal energy as a thermodynamic parameter is limited by the fact that it is more convenient (and perhaps safer!) to work at constant pressure than at constant volume. Where any expansion or contraction takes place, the system is doing work on its surroundings, or is having work done on it, so that the value of U will change even though no heat has passed in or out and no conscious effort has been made to do work on the system.

Consequently, we need to introduce the enthalpy, denoted by H and defined as:

$$H = U + PV \tag{6.4}$$

Let us assume that in going from an initial to a final state, the system, kept at constant pressure, P, expands by an amount ΔV. We then have:

$$\Delta H = \Delta U + P.\Delta V$$
$$= q + w + P.\Delta V \tag{6.5}$$

If the work involved is entirely related to expansion or contraction, then $w = -P.\Delta V$, so that ΔH is equal to q, the net quantity of heat passed into the system.

6.3 Calorimetry

When a chemical reaction takes place, there is, in general, a need either to remove heat from the system, or to pass heat into the system in order to bring the reaction products to the same temperature as the reactants were at initially. The experimental study of the amounts of heat involved is known as calorimetry.

It is sufficient here to describe one technique of calorimetry. In this, the reacting species (kept separate) are contained within a vessel and constitute the system under study. The vessel is so constructed and arranged that, as far as is possible, no heat is passed into the system: the technical term for this is *adiabatic*.

An illustration of an adiabatic calorimeter is shown in Figure 6.1, labelled as for the study of the thermochemistry of a reaction of an acid, A, with a base, B. The containing vessel is a Dewar flask and the holes in the bung are made no larger than is needed for operating the calorimeter. Inevitably, some heat will pass between the calorimeter and the surroundings. The aim is to assess the hypothetical temperature change, consequent on the chemical reaction, that would occur if the system were totally adiabatic.

After the calorimeter has been assembled and charged, the solution should be stirred at a steady rate while the temperature is recorded at regular time intervals. Then the solutions containing the two reactants are mixed and the stirring is continued as is the recording of the temperature. The stirring ensures that homogeneity is achieved and it is continued throughout so that whatever effect it has on the solution temperature will be a constant influence. A typical plot of temperature against time for an experiment of this nature is shown in Figure 6.2.

In general, in a calorimeter the species A and B react together to yield C. This is achieved adiabatically, with $q = 0$, and the temperature of our system has risen by ΔT. If we can now remove energy from the system, so that the temperature is restored to the original value, then the overall process will be one in which reactants A plus B at a certain temperature have yielded the product C at the same temperature. The enthalpy change for this will then be the standard enthalpy change, $\Delta H^{\ominus}$, for that amount of reaction:

$$\Delta H^{\ominus} = -C_p \text{ (products + calorimeter)}.\Delta T \tag{6.6}$$

The calorimetry experiment has yielded ΔT. To evaluate $\Delta H^{\ominus}$ for whatever amount of reaction has occurred, we need to know the heat capacity, C_p, of the reaction products and the calorimeter. To establish this may require another

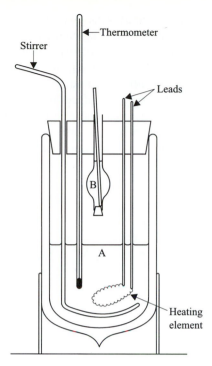

Figure 6.1 An illustration of an adiabatic calorimeter, with the reactants A and B, initially kept separate.

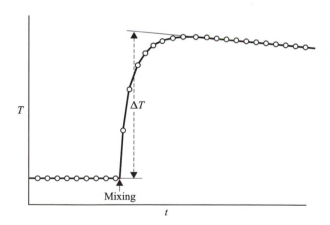

Figure 6.2 A typical plot of temperature (T) against time (t) for an adiabatic calorimeter, illustrating how the temperature change, ΔT, can be evaluated. The lines showing the trends of T before and after mixing are extrapolated linearly. To estimate ΔT, allowing for the heat gain or loss by the calorimeter, a vertical line is then drawn so that the two 'triangles' between this line and the actual and extrapolated lines for ΔT have equal areas.

experiment, in which the temperature rise is measured when a known amount of electrical work is done on the heating element within the calorimeter.

One practical point is that calorimetric experiments have significance only if one knows the extent of reaction occurring. Ideally, the reaction should go at least 99.99% of the way to completion, and this stage should be achieved quite quickly. The location of the position of equilibrium in chemical reactions is discussed in Chapter 8. For reactions where equilibrium does not lie so far over to the product side, it may be helpful to add an excess of one reagent so that one can aim to achieve almost total reaction of the other one.

— EXAMPLE 6.1

When 200 cm^3 of 0.5 mol dm^{-3} ethanoic acid is mixed in a calorimeter with 200 cm^3 of 0.5 mol dm^{-3} NaOH solution, ΔT is measured as 2.75 K. Given that the heat capacity of 0.25 mol dm^{-3} sodium ethanoate is 4.17 J K^{-1} cm^{-3}, and that the effective heat capacity of the calorimeter is 155 J K^{-1}, evaluate the molar standard enthalpy change for the neutralisation reaction.

— SOLUTION

The total heat capacity of the products plus calorimeter is given by:

$$C_p = (400 \text{ cm}^3 \times 4.17 \text{ J K}^{-1} \text{ cm}^{-3}) + 155 \text{ J K}^{-1}$$
$$= 1823 \text{ J K}^{-1}$$

Thus, the standard enthalpy change for the amount of reaction occurring is

$$\Delta H^{\ominus} = -1823 \text{ J K}^{-1} \times 2.75 \text{ K}$$
$$= -5013 \text{ J}$$

To obtain the value per mole, we divide this by the number of moles of acid being neutralised:

$$\Delta H^{\ominus} = \frac{-5013 \text{ J}}{(0.2 \text{ dm}^3 \times 0.5 \text{ mol dm}^{-3})}$$
$$= -50.1 \text{ kJ mol}^{-1}$$

As the solution to Example 6.1 illustrates, the standard enthalpy change of a reaction which occurs readily is frequently (but not invariably) negative. This arises from the manner in which H and ΔH are defined.

6.4 Hess's Law

If the standard enthalpy changes of two reactions have been measured, then these reactions may be combined to yield a third one, which may perhaps be a totally hypothetical process. For this new reaction, the standard enthalpy change may be found by combining the standard enthalpy changes for the two original reactions, just as their chemical equations have been combined. This, in essence, is Hess's Law, which does not contain any new principle that is not implicit in the First Law of Thermodynamics.

The standard enthalpy of formation of a compound is defined as the standard enthalpy change of the reaction (frequently, one which could not be caused to take place!) in which one mole of this compound is formed from its elements in their standard states. This establishes, as a reference point, the form of each element present at one atmosphere pressure and 298 K (or, if there is more than one, the more stable form). So the standard enthalpy of formation of $N_2(g)$ or of $O_2(g)$ or of C(graphite) is zero by definition, whereas that of $O_3(g)$ or of C(diamond) is non-zero.

Organic compounds containing the elements C, H and perhaps O are readily combusted in excess oxygen to give gaseous CO_2 and liquid water, and these reactions are readily amenable to calorimetric study. From such calorimetric studies it is known that the standard enthalpy of combustion of ethanol, $C_2H_5OH(l)$, is -1366.8 kJ mol^{-1} and those of C(graphite) and $H_2(g)$ are -393.5 and -285.9 kJ mol^{-1}. We may use these data to evaluate the standard enthalpy of formation of ethanol and this calculation exemplifies the use of Hess's Law.

In doing calculations of this type, errors may arise because the reactions used, while involving the substances required, also involve other substances of which the student has not accurately kept track. These may be averted if one always writes balanced chemical equations and, as these are added or subtracted, their standard enthalpy changes are treated in the same way. The data given may be written in equation form as follows:

(I) $C_2H_5OH(l) + 3O_2(g) = 2CO_2(g) + 3H_2O(l)$
$$\Delta H^{\ominus} = -1366.8 \text{ kJ mol}^{-1}$$

(II)   C(graphite) + $O_2(g) = CO_2(g)$ $\Delta H^{\ominus} = -393.5 \text{ kJ mol}^{-1}$

(III) $H_2(g) + \frac{1}{2}O_2(g) = H_2O(l)$ $\Delta H^{\ominus} = -285.9 \text{ kJ mol}^{-1}$

The reaction for which one wishes to discover the standard enthalpy change is:

   2C(graphite) + $3H_2(g) + \frac{1}{2}O_2(g) = C_2H_5OH(l)$

To obtain this, one should take the reverse of Equation I, add twice Equation II and three times Equation III:

$$\Delta H^{\ominus}/\text{kJ mol}^{-1}$$

$2CO_2(g) + 3H_2O(l)$	$= C_2H_5OH(l) + 3O_2(g)$	$+1366.8$
$2C(\text{graphite}) + 2O_2(g)$	$= 2CO_2(g)$	-787.0
$3H_2(g) + 1\frac{1}{2}O_2(g)$	$= 3H_2O(l)$	-857.7

$$2C(\text{graphite}) + 3H_2(g) + O_2(g) = C_2H_5OH(l) \qquad -277.9$$

So the standard enthalpy of formation of ethanol has been determined. This refers, of course, to a reaction, the quantitative conversion of graphite to ethanol in a single process, which one could hardly hope to achieve in the laboratory.

In reference books such as the *Handbook of Chemistry and Physics* (CRC Press) one finds listed both the standard enthalpies of combustion of various organic compounds and their standard enthalpies of formation. Data from either of these lists may be used to evaluate the standard enthalpy changes for other chemical reactions in which such compounds are either reactants or products.

6.5 The dependence of the standard enthalpy change on temperature

Tabulated values of the standard enthalpies of combustion or of formation normally refer, on grounds of consistency, to the common temperature of 298 K. Values of $\Delta H^{\ominus}$ are not independent of temperature, but the extent of their variation with temperature is usually slight.

Consider the hypothetical reaction,

$$A + B = C + D \qquad\qquad (6.7)$$

which we may conceive of as taking place either at temperature T_1, or at the higher temperature, T_2.

$$T_2 \qquad A + B \qquad \rightarrow \qquad C + D$$

$$\downarrow \qquad\qquad\qquad \uparrow$$

$$T_1 \qquad A + B \qquad \rightarrow \qquad C + D$$

For the process in which we start with A + B at temperature T_2 and finish with C + D at the same temperature, we may envisage two alternative routes. One is the direct route, for which the standard enthalpy change will be written, $\Delta H^{\ominus}$ (T_2). The other is the three step route, where the reactants, A + B, are first cooled down to T_1, at which temperature the reaction takes place to give C + D. The requisite amount of heat is then supplied to bring the temperature of the products up to T_2. By the First Law of Thermodynamics, the two enthalpy changes must be equal, so we have:

$$\Delta H^{\ominus}(T_2) = -C_p(A + B).(T_2 - T_1) + \Delta H^{\ominus}(T_1) + C_p(C + D)(T_2 - T_1)$$
$$= \Delta H^{\ominus}(T_1) + \{C_p(C + D) - C_p(A + B)\}(T_2 - T_1)$$

$$(6.8)$$

where $C_p(A + B)$ and $C_p(C + D)$ denote the heat capacities at constant pressure of the reactants and the products. Where these are appreciably different, the last term of equation (6.8) will be sufficiently large that it should not be neglected. If the temperature difference, $(T_2 - T_1)$ is very large then this last term may become appreciable. However, heat capacities are not independent of temperature and over a large temperature range this variation cannot be ignored: it would then be desirable that in the first line of Equation (6.8), the appropriate integrals are used for the heat removed from the reactants and for that added to the products. Equation (6.8) is known as Kirchhoff's equation.

6.6 Bond enthalpies

We may conceive of a reaction in which we start with a molecule in the gas phase and finish with its individual component atoms, all also in the gas phase. For example:

$$CH_4(g) = C(g) + 4H(g) \tag{6.9}$$

Energy needs to be expended to achieve this dismemberment of the methane molecule, so $\Delta H^{\ominus}$ for this process will necessarily be positive. Since the methane molecule contains identical chemical bonds, if we now divide this standard enthalpy change by the number of the identical bonds that have been broken, we will have derived the mean bond enthalpy of the C–H bond in methane.

The unwary may think that for Reaction (6.9), the standard enthalpy change should be the reverse of the standard enthalpy of formation of the compound. This is a misguided notion. Such people have failed to notice that, while the species on the right-hand side of this equation are the constituent elements, these elements are *not* in their standard states. This distinction is a crucial one.

—— **EXAMPLE 6.2** ————————————————————————
Given that the standard enthalpy of combustion of $CH_4(g)$ is -890.5 kJ mol^{-1}, of $H_2(g)$ is -285.9 kJ mol^{-1} and of C(graphite) is -393.5 kJ mol^{-1}, that the bond dissociation enthalpy of $H_2(g)$ is 435.8 kJ mol^{-1} and the latent heat of sublimation of graphite is 718.4 kJ mol^{-1}, evaluate the bond mean enthalpy of the C–H bond in methane.

— SOLUTION —————————————————————————————

The first objective is to find the standard enthalpy change for Equation (6.9), which is a Hess's Law problem:

$$\Delta H^{\ominus} / kJ\ mol^{-1}$$

$CH_4(g) + 2O_2(g) = CO_2(g) + 2H_2O(l)$		−890.5
$CO_2(g)$	$= C(gr) + O_2(g)$	+393.5
$2H_2O(l)$	$= 2H_2(g) + O_2(g)$	+571.8
$2H_2(g)$	$= 4H(g)$	+871.6
$C(gr)$	$= C(g)$	+718.4

$CH_4(g)$	$= C(g) + 4H(g)$	+1664.8

Thus, the mean bond enthalpy of the C–H bond, H(C–H), in CH_4 is 416.2 kJ mol^{-1}.

———

For the vast majority of compounds, the chemical bonds are not all identical. For these, also, the energy required to atomise the gaseous molecule must be equal to the aggregate of the bond enthalpies. The problem is to discover in what way this total should be shared out between the various bonds.

More detailed studies suffice to show that the bond enthalpy of a particular bond depends on its chemical environment. For the C–H bond within alkanes, it is easily shown that the value must depend on the degree of substitution of the carbon atom, becoming progressively less the more highly substituted it is. Thus the bond enthalpy of the $(CH_3)_3C–H$ bond in 2-methylpropane is estimated as appreciably less than the figure calculated above for methane.

Some bond enthalpy values are quoted in Table 6.1. Those for diatomic and symmetrical polyatomic molecules are obviously more reliable than those relating to molecules with several different types of chemical bond. However the latter group, although bearing this general health warning, give helpful indications of the strengths of these chemical bonds.

6.7 The Second Law and the entropy

There are many possible ways in which the Second Law of Thermodynamics may be stated, and they all express the same fundamental law of nature. To an engineer, the Law has important consequences for the efficiency of a heat engine and it may be expressed succinctly in such terms. To a chemist, a

Table 6.1   Bond enthalpies of selected chemical bonds

Chemical bond	Bond enthalpy/kJ mol^{-1}
H–H	436
O=O	497
N≡N	959
Li–Li	103
Na–Na	77
K–K	54
F–F	157
Cl–Cl	243
Br–Br	194
I–I	153
H–F	569
H–Cl	431
H–Br	366
H–I	299
C=O (in CO)	1076
C=O (in CO$_2$)	804
C=S (in CS$_2$)	578
N–H (in NH$_3$)	391
O–H (in H$_2$O)	463
C–H (in CH$_4$)	416
C–F (in CF$_4$)	428
C–Cl (in CCl$_4$)	327
C–Br (in CBr$_4$)	263
⋝C–C⋜	368
⋝C=C⋜	710
–C≡C–	962

The above data have mostly been calculated from the literature values for the enthalpies of formation.

more important aim relates to the conditions for chemical or physicochemical equilibrium. In this, the concept of entropy is vital and so, from a chemist's perspective, the role of the Second Law is to define what is meant by a change in entropy.

First, it will be useful to explain the scientific use of the word 'reversible'. A *reversible* compression, for example, is one performed in such a way that an infinitesimal change in the applied pressure would reverse the direction of the volume change. Passing in heat *reversibly* means it is done in such a way that if the temperature of the heat sink were decreased infinitesimally, the direction of heat flow would be reversed. This means that a reversible change of such a nature must take place extremely slowly. However, some other processes are intrinsically irreversible, regardless of their rapidity. The mixing of two gases cannot be a reversible process, irrespective of what constraints we may impose to slow the process: we simply cannot conceive of the mixing process being reversed to give back the original pure gases.

We can now express the Second Law in two simple statements:

(a) For any closed system, taken from state 1 to state 2 by a reversible path, the quotient q_{rev}/T is constant, regardless of path. Where the heat transfer takes place at different temperatures, we require the sum of such quotients, $\Sigma(q_{rev}/T)$, and if T is continually varying then we require the integral, $\int(dq_{rev}/T)$.

(b) For any irreversible path between states 1 and 2, q/T is variable but is always less than q_{rev}/T.

These statements may be seen in the context of the First Law, which states that $(q + w)$ is constant. If the heat is passed in reversibly at a constant temperature, q_{rev}/T is invariant and so it has all the properties of a function of state. In effect, the Second Law defines a change in the entropy, S:

$$\Delta S = S_2 - S_1 = \frac{q_{rev}}{T} \tag{6.10}$$

The reason, in general, for the lower value for the quotient where the process is irreversible is that T has been greater than it need have been, not that any less heat is required.

A simple example of an entropy change is provided by the entropy of fusion of ice, where we start with ice at 273 K and finish with liquid water at the same temperature. The latent heat of fusion is 334.7 J g^{-1}, so for one mole, we have:

$$\Delta S_{fus} = \frac{334.7 \text{ J g}^{-1} \times 18.0 \text{ g mol}^{-1}}{273 \text{ K}}$$
$$= 22.06 \text{ J K}^{-1} \text{ mol}^{-1} \tag{6.11}$$

In passing heat into a substance in a single phase, if the heat capacity C_p is constant then we have:

$$\Delta S = S_2 - S_1 = \int_1^2 \frac{C_p dT}{T}$$
$$= C_p \ln(T_2/T_1) \tag{6.12}$$

Suppose we start with 100 g of water at 283 K, and pass heat into it until a temperature of 363 K is attained. If the heat capacity of water is assumed constant at 4.2 J g^{-1} K^{-1}, then the entropy increase is given by:

$$\Delta S = 4.2 \text{ J g}^{-1} \text{ K}^{-1} \times 100 \text{ g} \times \ln\frac{363}{283}$$
$$= 104.6 \text{ J K}^{-1} \tag{6.13}$$

We can now contemplate evaluating the entropy change for an irreversible process, namely that of mixing adiabatically two 100 g portions of water, one at 283 K and the other at 363 K. The essence of our approach is that we will devise a reversible path, admittedly a totally hypothetical one, to take us from the initial state to the final state. We first calculate that the final temperature, after mixing, is 323 K.

Our path is first to pass heat reversibly into the 100 g of cold water until it attains 323 K, then to remove heat reversibly from the hot water until it attains the same temperature, and finally to mix the two. The total entropy change will be the sum of the contributions from these three steps.

$$\Delta S = \Delta S_1 + \Delta S_2 + \Delta S_3$$

$$= (4.2 \times 100) \text{ J K}^{-1} \ln \frac{323}{283} + (4.2 \times 100) \text{ J K}^{-1} \ln \frac{323}{363} + 0 \qquad (6.14)$$

$$= (55.5 - 49.0) \text{ J K}^{-1}$$

$$= +6.5 \text{ J K}^{-1}$$

For this process, as it was performed, q_{irrev} was zero, as would be q_{irrev}/T. The sign of our answer illustrates statement (b) above.

6.8   The requirements for physicochemical equilibrium

Consider the system composed of two bulbs, containing two different, non-reacting gases at the same temperature and pressure and separated by a closed tap. When the tap is opened, the gases will mix. This will bring about no change in the pressure or the temperature or the enthalpy of the system. However, it is a spontaneous or irreversible rather than a reversible process, so the entropy, S, will increase.

The German physicist Clausius summed up his version of the First and the Second Laws of Thermodynamics in two memorable sentences. In translation, they read:

The energy of the universe is constant.
The entropy of the universe rises to a maximum.

The mixing of the two gases, referred to above, supports the second of these assertions. Under the particular circumstances applicable here, where the volume is kept constant, the only change that may occur in the system is one where the entropy increases.

In general, when chemical reactions may occur, there is a tendency to form the strongest chemical bonds. Thus, a mixture of H_2 and F_2 is almost quantitatively converted into $2HF$, in which the aggregate of the bond enthalpies of the molecules has clearly increased. It was once (but erroneously) proposed that the only chemical reactions that may take place are those with a negative standard enthalpy change. However, the fallacy of this statement is attested by the rare but memorable examples of chemical reactions with positive standard enthalpy changes, but also positive and large entropy changes. The real question is, how can we predict the direction and the extent of change in the general situation where both the enthalpy change and the entropy change are non-zero?

In a purely mechanical system, such as a ball-bearing free to move along a groove, the criterion of equilibrium is that the potential energy should be minimised. So the ball-bearing comes to rest at the lowest accessible point along the groove. We would like to have a thermodynamic function which could play a similar role for physicochemical systems in which chemical reactions and phase changes may occur.

However, it is necessary to refine the question before an answer can be given. One must first specify what are the constraints. In the laboratory, we usually work at atmospheric pressure and at constant temperature. These parameters, P and T, are called intensive variables because they are independent of the amount of substance. So our major question is: how can we predict what will happen at constant temperature and pressure? Very occasionally, instead of controlling the pressure we may keep the volume constant: our requirements then relate to constant temperature and volume.

Let us define two additional thermodynamic functions, the Helmholtz energy, A, and the Gibbs energy, G:

$$A = U - TS \tag{6.15}$$
$$G = H - TS$$
$$= U + PV - TS \tag{6.16}$$

From this last equation, we may express a small change in G as follows:

$$dG = dU + P.dV + V.dP - T.dS - S.dT$$
$$= q + w + P.dV + V.dP - T.dS - S.dT \tag{6.17}$$

This last line has been attained by substituting from Equation (6.3), and thus incorporating the First Law. In Equation (6.5), it was shown that w cancels with $P.dV$ provided the work involved is entirely related to expansion or contraction. Also, by invoking the Second Law, Equation (6.10), q cancels with $-T.dS$. Thus we have,

$$dG = V.dP - S.dT \tag{6.17a}$$

which shows that, at constant pressure and temperature, there should be no change in the Gibbs energy of the system. Further analysis shows that, under these conditions, G will attain a minimum value when the system is at equilibrium.

In a similar way, one can deduce the equation,

$$dA = P.dV - S.dT \tag{6.18}$$

which shows that V and T are the natural variables for the Helmholtz energy, A. This means that at constant volume and constant temperature, the Helmholtz energy of the system is a minimum.

Suggested reading

SMITH, E. B., 1990, *Basic Chemical Thermodynamics*, 4th Edn, London: Oxford University Press.

WARN, J. R. W., 1976, *Concise Chemical Thermodynamics*, London: Chapman & Hall.

WASER, J., 1966, *Basic Chemical Thermodynamics*, New York: Benjamin.

McGLASHAN, M. L., 1966, The uses and abuses of the laws of thermodynamics, *Journal of Chemical Education*, **43**, 226–32.

BENSON, S. W., 1965, Bond Energies, *Journal of Chemical Education*, **42**, 502–18.

Problems

6.1 Evaluate how much heat is required to convert 5 kg of ice at 273 K into steam at 373 K, all at one atmosphere pressure, given that the latent heat of fusion of ice is 333.5 kJ kg^{-1}, the latent heat of vaporisation of water is 2257.3 kJ kg^{-1} and the heat capacity of liquid water is 4190 J K^{-1} kg^{-1}.

6.2 The standard enthalpy of combustion of phthalic acid, $C_6H_4(CO_2H)_2$, is -3223.5 kJ mol^{-1}. Evaluate its standard enthalpy of formation, given as follows:

$$\Delta H_f^{\ominus}(CO_2, g) = -393.5 \text{ kJ mol}^{-1}$$
$$\Delta H_f^{\ominus}(H_2O, l) = -285.9 \text{ kJ mol}^{-1}$$

6.3 The standard enthalpy of combustion of 2-propanol(l) is -1986.6 kJ mol^{-1}. Evaluate the standard enthalpy of formation of this compound, given that the standard enthalpies of formation of $H_2O(l)$ and $CO_2(g)$ are respectively -285.9 and -393.5 kJ mol^{-1}.

Given also that the standard enthalpy of combustion of 2-propanone is -1790.4 kJ mol^{-1}, evaluate the standard enthalpy change for the hydrogenation (i.e., reaction with H_2) of 2-propanone to 2-propanol.

6.4 The standard enthalpy of combustion of glucose at 298 K is -2802 kJ mol^{-1}. For a man of 80 kg to climb a mountain with a vertical ascent of 2000 m, calculate how much glucose would need to be metabolised, assuming that 25% of the energy released can be used in muscular work.

At the summit of the mountain, the climber was soaked and a cold wind dried 1 kg of water from his clothing. How much glucose would need to be metabolised to make good the heat loss? If no glucose were available, by how much would his body temperature be lowered, assuming the heat capacity of a human body equals that of an equivalent mass of water.

Latent heat of vaporisation of water = 2445 J g^{-1}
Heat capacity of water = 4.2 J K^{-1} g^{-1}
Acceleration due to gravity, g = 9.81 m s^{-2}
M.W. of glucose, $C_6H_{12}O_6$ = 180 g mol^{-1}
Work in climbing = Mass × (Vertical ascent) × g

6.5 Estimate the standard enthalpy change for the oxidation (by O_2) of lactic acid, $CH_3CH(OH)CO_2H$ to pyruvic acid, CH_3COCO_2H, given the following standard enthalpies of formation:

$\Delta H_f^{\ominus}$ (pyruvic acid, l) $= -584.5$ kJ mol^{-1}
$\Delta H_f^{\ominus}$ (lactic acid, s) $= -674.2$ kJ mol^{-1}

Any necessary additional data have been supplied in earlier questions.

6.6   Calculate the mean bond enthalpy of the O–O bond in ozone (O_3), given that the standard enthalpy of formation of $O_3(g)$ is $+142.3$ kJ mol^{-1} and that the bond dissociation enthalpy of O_2 is 497.3 kJ mol^{-1}. (NB. The ozone molecule is bent, with a bond angle of 117°.)

6.7   Evaluate the mean bond enthalpy in silane (SiH_4) given that the standard enthalpy of formation of $SiH_4(g)$ is -61.5 kJ mol^{-1}, the molar enthalpy of sublimation of silicon is 368.4 kJ mol^{-1} and the standard enthalpy of dissociation of $H_2(g)$ is 435.8 kJ mol^{-1}. (NB. In its standard state, silicon has the same phase as does carbon.)

6.8   Evaluate the standard enthalpy change at 298 K for the reaction,

$$SOCl_2(l) + H_2O(l) = SO_2(g) + 2HCl(g)$$

given the following standard enthalpies of formation, all at 298 K:

$SOCl_2(l)$	-205.9 kJ mol^{-1}
$H_2O(l)$	-285.9 kJ mol^{-1}
$SO_2(g)$	-296.9 kJ mol^{-1}
$HCl(g)$	-92.3 kJ mol^{-1}

Also, evaluate the standard enthalpy change of the same reaction at 350 K, given the following molar heat capacities:

$SOCl_2(l)$	114.5 J K^{-1} mol^{-1}
$H_2O(l)$	75.4 J K^{-1} mol^{-1}
$SO_2(g)$	40.7 J K^{-1} mol^{-1}
$HCl(g)$	29.6 J K^{-1} mol^{-1}

6.9   Given the following data,

	Argon	n-Butane	2-Propanol
Boiling point/°C	-185.7	-0.5	82.4
L_{vap}/J g^{-1}	157.3	382.8	665.3
MW/g mol^{-1}	39.9	58.1	60.1

evaluate which of these three substances has the largest molar entropy of vaporisation at its boiling point.

6.10   Evaluate the entropy change when 1 kg of ice at 273 K is added adiabatically to 1 kg of water at 370 K. Use the relevant data supplied in problem 6.1.

7

Phase equilibria and solutions of non-electrolytes

The simplest phase equilibria are those involving a single substance. If one phase is the vapour, then the manner in which the vapour pressure rises with temperature reflects the latent heat involved in the phase change.

For mixtures of two liquids, the norm of ideal behaviour is specified on the basis of the partial pressure being proportional to the mole fraction, for each component. Where this is obeyed, the two liquids may be separated by fractional distillation. Also, where solutions behave ideally, various properties serve as useful probes of the state of aggregation (or of dissociation) of the solutes in solution.

Immiscibility of two liquids is a consequence of substantial differences in the respective intermolecular forces. Solutes distribute themselves between the two in a way which reflects their interactions with both liquids, and in accordance with the general principle that the Gibbs energy of the system should be minimised.

7.1 The Clapeyron equation

In considering equilibria between phases of a single substance, or transitions between such phases, it is useful to start with the most basic equation. Suppose we have two phases of a substance, which we will label α and β. Equation (6.17a) applies to both phases, so we have:

$$\left. \begin{array}{l} dG_\alpha = V_\alpha.dP - S_\alpha.dT \\ dG_\beta = V_\beta.dP - S_\beta.dT \end{array} \right\} \qquad (7.1)$$

These two phases are, of course, at the same temperature, T, and pressure, P, and since they are in equilibrium we may equate the infinitesimal changes of the Gibbs energies of the two phases to obtain:

$$V_\alpha.dP - S_\alpha.dT = V_\beta.dP - S_\beta.dT \qquad (7.2)$$

Rearranging, we have:

$$\frac{dP}{dT} = \frac{S_\beta - S_\alpha}{V_\beta - V_\alpha} = \frac{\Delta S}{\Delta V} \qquad (7.3)$$

On the right-hand side of this equation, the Δ signs refer to the difference between the two phases, both in the same sense. In evaluating the ratio, $\Delta S/\Delta V$, one must take the volume change and the entropy change for the same amount of substance. Equation (7.3) is widely known as the Clapeyron equation and applies to any two phases.

In a phase diagram such as Figure 5.4, the ordinate is P and the abscissa T. Thus the left-hand side of Equation (7.3) is the slope of a line in a phase diagram denoting the coexistence of two phases. The equation says that this is equal to the ratio of the change in entropy to the change in volume accompanying the phase transition.

For any substance not subjected to extreme pressures, the vapour phase occupies a much greater volume than does the liquid or the solid and, in the light of the substantial latent heats of vaporisation and of sublimation, the entropy of a vapour is greater, at the same temperature, than that of the liquid or the solid. So, for both solid–vapour and liquid–vapour equilibria, $\Delta S/\Delta V$ is positive. We have already commented in section 5.8, in relation to Figure 5.4, that for solid–vapour and liquid–vapour equilibria, the vapour pressure increases with the temperature so that dP/dT is positive.

Regarding solid–liquid equilibria, there is always a positive latent heat of fusion, which attests to an increase in the entropy. The volume change is always much less than for either of the previous transitions. Usually, V increases on melting, because almost invariably fusion is accompanied by a slight expansion, so for solid–liquid equilibria, $\Delta S/\Delta V$ is usually large and positive. However, for a very small number of substances, fusion is accompanied by a slight contraction, so that $\Delta S/\Delta V$ is large and negative.

7.2 The anomalous behaviour of water

The most notable of these few substances is water, which has an anomalously low density in the solid state. This arises, ostensibly, because the hydrogen-bonded structure of ice means that each molecule is co-ordinated to only four others, as depicted in Figure 5.6.

When ice melts and the tetrahedral array is broken up there is a substantial contraction, but considerable hydrogen-bonding persists. As the temperature is increased, the extent of the hydrogen-bonding is progressively diminished. Near to the melting point, the break-up of the hydrogen-bonding has more influence on the volume than has the increasing velocity of the molecules, which normally causes a liquid to expand with increasing temperature. Water, unusually, shows a maximum in its density, which occurs at 4°C.

The phase diagram of water, sketched in Figure 7.1, shows the negative slope of the solid–liquid coexistence line. In consequence, if the temperature

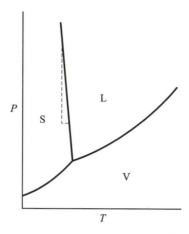

Figure 7.1 The phase diagram of water, indicating the phases present on a plot of P against T. The dotted line shows how a temperature below 0°C can generate a huge pressure or, alternatively, how a huge pressure can cause ice to melt below 0°C.

of liquid water is lowered below the freezing point, it may be seen that, with the volume restricted, the pressure must rise sharply. This demonstrates the thermodynamic reason for the bursting of water pipes in times of frost. If we used virtually any other liquid of a similar freezing point, a cold snap would hold no terrors but the thaw would be potentially disastrous.

Another aspect is exemplified if one starts with ice at a few degrees below the freezing point. By applying a considerable pressure to it, we can attain the solid–liquid coexistence line at which melting will be possible. In effect, this is what happens on an ice rink, where the weight of the skater is concentrated on the minuscule area of the blades of the skates. The resulting pressure causes the ice to melt below the blades, thus facilitating movement. In this instance, the anomalous behaviour of water is seen to advantage: scarcely any other substance, if used in an ice rink, could so facilitate the talents of ice skaters.

7.3 The Clausius–Clapeyron equation

The Clapeyron equation (7.3) is of general utility and is applicable to all phase transitions, including solid–solid transitions such as graphite–diamond. For equilibria involving a vapour phase, i.e. for solid–vapour and liquid–vapour equilibria, some simplifications are possible, leading to a more specific equation as to how the vapour pressure of a solid or a liquid should vary with temperature.

The first point is that, except under huge pressures, the volume of the vapour is so much greater than that of a condensed phase that, to an acceptable approximation, for ΔV we may substitute the volume of the vapour:

$$\Delta V = V_{vap} - V_{cp} \approx V_{vap} \tag{7.4}$$

Secondly, the behaviour of the vapour will not much differ from that of an ideal gas, and for one mole of the latter, $PV = RT$. So instead of V_{vap} we may substitute RT/P. We also require an expression for the entropy change for one mole, and this is given by the ratio of the molar latent heat to the Kelvin temperature. So we may substitute into Equation (7.3) as follows,

$$\frac{dP}{dT} = \frac{\Delta S}{\Delta V} = \frac{P}{RT} \cdot \frac{L}{T} \tag{7.5}$$

where L denotes the molar latent heat of sublimation or of vaporisation. The next step involves the mathematical manipulation called integration, which is outlined in Appendix Two. Before we can integrate this equation we need to separate the variables P and T:

$$\frac{dP}{P} = \frac{L}{R} \cdot \frac{dT}{T^2} \tag{7.6}$$

This gives, on integration:

$$\ln P = -\frac{L}{R} \cdot \frac{1}{T} + const \tag{7.7}$$

which is known as the Clausius–Clapeyron equation. This indicates that the logarithm of the vapour pressure should be a linear function of the reciprocal temperature and that the (negative) slope of this line is proportional to the latent heat of the relevant phase transition. So, measurements of the variation with temperature of the vapour pressure of a liquid can serve as a non-calorimetric route to measuring the latent heat of vaporisation.

Equation (7.7), however, is not an exact relationship, quite apart from the unknown constant of integration. In part, this is because of the approximations introduced in equation (7.5), but also it arises because Equation (7.6) has been integrated as if L were a constant. Strictly speaking, latent heats of vaporisation are temperature dependent and decrease as the temperature rises, eventually falling to zero at the critical temperature. For these reasons, a plot of $\ln P$ against T^{-1} is not strictly linear, but $-R$ times the slope at any point gives the latent heat applicable at that temperature.

—— **EXAMPLE 7.1** ——————————————————————————
The vapour pressure of H_2O varies with temperature as follows:

t/°C	p/torr	t/°C	p/torr
−25	0.476	5	6.54
−20	0.776	10	9.21
−15	1.241	15	12.79
−10	1.95	20	17.54
−5	3.01	25	23.76
0	4.58	30	31.82

Thus evaluate:

(a) the molar enthalpy of vaporisation of water,

(b) the molar enthalpy of sublimation of ice,

(c) the molar enthalpy of fusion of ice, and

(d) the anticipated boiling point of water, based on the above data.

— SOLUTION —

This is a more complex example than some, in that there are two different stable phases over the temperature range involved, and so there are two separate phase transitions, namely solid–vapour over −25 to 0°C and liquid–vapour over 0 to 30°C. Two different latent heats are involved over these temperature ranges, namely L_{sub} for the sublimation process and L_{vap} for the second. So when ln P is plotted against T^{-1}, we expect to have linear portions over each of these temperature ranges, but with different slopes.

$t/°C$	T/K	$T^{-1}/10^{-3}K^{-1}$	$P/torr$	ln $(P/torr)$
−25	248.1	4.030	0.476	−0.74
−20	253.1	3.950	0.776	−0.25
−15	258.1	3.874	1.241	0.22
−10	263.1	3.800	1.95	0.67
−5	268.1	3.729	3.01	1.10
0	273.1	3.661	4.58	1.52
5	278.1	3.595	6.54	1.88
10	283.1	3.532	9.21	2.22
15	288.1	3.470	12.79	2.55
20	293.1	3.411	17.54	2.86
25	298.1	3.354	23.76	3.17
30	303.1	3.299	31.82	3.46

The plot of ln P against T^{-1} is shown in Figure 7.2.

From the right-hand portion, pertaining to the lower temperature region, we have:

$$\text{Slope} = \frac{-0.89 - 1.58}{(4.05 - 3.65) \times 10^{-3} \text{ K}^{-1}} \tag{7.8}$$

$$= -6.17 \times 10^3 \text{ K}$$

$$\therefore L_{sub} = -8.314 \text{ J K}^{-1} \text{ mol}^{-1} \times -6.17 \times 10^3 \text{ K}$$

$$= 51.3 \times 10^3 \text{ J mol}^{-1} \tag{7.9}$$

$$= 51.3 \text{ kJ mol}^{-1}$$

This is the latent heat of sublimation of ice, and the answer to part (a).

From the left-hand portion, referring to the higher temperatures, we have:

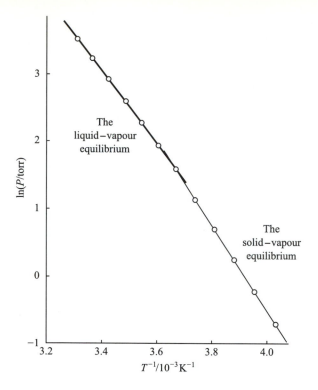

Figure 7.2 The plot of ln P against T^{-1} for the data in Example 7.1. The straight line on the right pertains to the equilibrium between ice and water vapour and the one on the left to that between liquid water and water vapour.

$$\text{Slope} = \frac{1.58 - 3.75}{(3.65 - 3.25) \times 10^{-3} \text{ K}^{-1}}$$

$$= -5.42 \times 10^3 \text{ K} \qquad (7.10)$$

$$\therefore L_{vap} = -8.314 \text{ J K}^{-1} \text{ mol}^{-1} \times -5.42 \times 10^3 \text{ K}$$

$$= 45.1 \times 10^3 \text{ J mol}^{-1} \qquad (7.11)$$

$$= 45.1 \text{ kJ mol}^{-1}$$

This is the latent heat of vaporisation of water and the answer to part (b). Another way of expressing these enthalpy changes is to write:

$$H_2O(s) = H_2O(g) \qquad \Delta H^{\ominus} = 51.3 \text{ kJ mol}^{-1}$$
$$H_2O(l) = H_2O(g) \qquad \Delta H^{\ominus} = 45.1 \text{ kJ mol}^{-1}$$

Subtracting these, we obtain,

$$H_2O(s) = H_2O(l) \qquad \Delta H^{\ominus} = 6.2 \text{ kJ mol}^{-1}$$

which yields the latent heat of fusion of ice and is the answer to part (c).

Part (d) invites us to extrapolate the liquid–vapour line in Figure 7.2 until it attains $P = 760$ torr or $\ln (P/\text{torr}) = 6.63$. We might redraw Figure 7.2, choosing scales that would permit this to be done. Alternatively, we may calculate the temperature where P would reach 760 torr, using our present line.

Knowing that the line has a slope of -5.42×10^3 K and that it passes through the point $(3.354 \times 10^{-3}$ K^{-1}, 3.17), we may write the equation for this line as

$$\ln (P/\text{torr}) = -5.42 \times 10^3 T^{-1} + 21.35 \tag{7.12}$$

Substituting for P, we have:

$$T^{-1} = \frac{21.35 - 6.63}{5.42 \times 10^3 \text{ K}}$$
$$= 2.716 \times 10^{-3} \text{ K}^{-1} \tag{7.13}$$

So $T = 368.2$ K, or 95.1°C. This shows that, by extrapolating our plot, the answer obtained for the boiling point is liable to be a few degrees in error.

Sometimes students worry over the units of pressure to be used in calculations of this kind, and wonder if it is necessary to use SI units. Suppose, to convert to SI units, all the pressures given need to be multiplied by the factor f. When the logarithms are taken, these are of $\ln (fP) = \ln f + \ln P$, which means that the ordinate of every point has been increased by the constant amount of $\ln f$. Clearly, this has no effect on the slope of the line, so the latent heat may be evaluated, as in Example 7.1, using any units for P.

7.4 The miscibility of liquids

Everyone knows that oil and water do not mix, and that when the two liquids are put into the same vessel, two layers are obtained. In effect, the more dense liquid is at the bottom and the less dense is at the top, but mixing always occurs to a finite extent, which in certain cases may be extremely slight.

As a general criterion, like mixes freely with like, so that alkanes mix freely with each other and the lower alcohols intermix freely. Also, the lower alcohols mix freely with water, since both contain the –OH group. Non-mixing arises only when the molecules of the two substances are quite different so that the intermolecular forces within one liquid are quite different from those within the other. In that case, the mixing process entails work being done in breaking up the intermolecular attractions within each liquid.

In thermodynamic terms, mixing two liquids while maintaining a constant total volume is always accompanied by an increase in entropy. Since $\Delta G = \Delta H - T.\Delta S$, this means that ΔG will be negative unless ΔH for the mixing process is substantially positive. Where mixing causes a total change in the intermolecular attractive forces, then ΔH for the mixing process may well be sufficiently positive so that mixing does not occur.

The most significant experimental variable in this regard is the temperature. If ΔH and ΔS are both independent of temperature, then, since ΔS is positive, an increase in temperature will inevitably cause a decrease in ΔG. There are examples of pairs of liquids which form two phases at certain temperatures but which mix freely at higher temperatures. The value at which complete misci- bility is achieved is called the upper critical solution temperature and the behaviour of one such system is illustrated in Figure 7.3. This does not mean that all pairs of liquids are fully miscible at higher temperatures, since the value of T at which $T.\Delta S$ would become sufficiently large in relation to ΔH may be far in excess of the boiling points of the two liquids.

An interesting type of behaviour is exhibited by certain substances of large molar mass which are insoluble in water, such as carboxylic acids with a large alkyl or alkenyl group. If, for example, a very small amount of a dilute pentane solution of oleic acid is put on to the surface of water, the pentane rapidly evaporates and the acid spreads over the surface. Where the amount of acid is not more than is necessary for a complete monolayer on the surface, then the carboxylic group of each molecule is immersed in the aqueous layer.

Even more pronounced alignment of large molecules with moieties of dif- ferent characteristics may be achieved at an interface between two immiscible liquids, of which one phase is usually an aqueous one. If the other layer is a

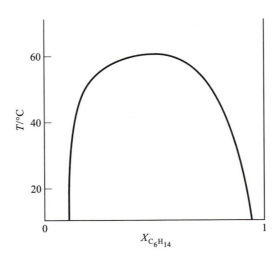

Figure 7.3 The phase diagram for the system aniline/*n*-hexane at constant pressure, showing how two phases are present up to 59.6°C, the upper critical solution temperature. (Data from Keyes and Hildebrand, *J. Am. Chem. Soc.*, 1917, **39**, 2297.)

hydrocarbon, then a higher alkanoic acid or a higher alcohol (such as 1-octanol), both of which are insoluble in water, will tend to gather at the interface, with the polar functional group poking into the aqueous layer where it can engage in hydrogen bonding and the alkyl group will be within the hydrocarbon layer, surrounded by molecules of like character.

7.5 Raoult's Law for binary liquid mixtures

Suppose a pair of liquids mix freely in all proportions. If we prepare a mixture of the two and put it into a closed vessel from which all air has been removed, it is of interest to know the pressure and composition of the vapour in equilibrium with the liquid. The fact that the two liquids freely mix precludes their being of utterly different character: the behaviour which is prescribed as par for such a mixture is that to be anticipated if the molecules of the two are virtually identical.

We need first to define the appropriate scale of composition for a binary mixture. This is expressed in terms of the fractions of the molecules present belonging to each component, called the mole fractions. For a liquid mixture of *A* and *B*, the mole fractions, X_A and X_B, of components A and B are given by,

$$\left. \begin{aligned} X_A &= \frac{n_A}{n_A + n_B} \\ X_B &= \frac{n_B}{n_A + n_B} \end{aligned} \right\} \tag{7.14}$$

where n_A and n_B denote the numbers of moles of each substance. Clearly, X_A and X_B add up to unity.

By Raoult's Law, if a binary liquid mixture behaves ideally, then the partial pressure of each component is proportional to its mole fraction. This means that the partial pressure of *A*, P_A, is the product of the mole fraction, X_A, and P_A^o, the vapour pressure of pure component *A* at the same temperature:

$$\left. \begin{aligned} P_A &= X_A P_A^o \\ P_B &= X_B P_B^o = (1 - X_A) P_B^o \end{aligned} \right\} \tag{7.15}$$

These equations are readily summarised diagrammatically as is shown in Figure 7.4. The total vapour pressure of the liquid mixture is $(P_A + P_B)$ and in the light of Equation (7.15), this is represented, as a function of composition, by the straight line from P_B^o at $X_A = 0$ to P_A^o at $X_A = 1$.

As to the composition of the vapour, in the light of Avogadro's hypothesis, the ratio of the numbers of molecules of the two species in a certain volume is the same as the ratio of their partial pressures. Using *y* to denote mole fractions in the vapour phase, we have:

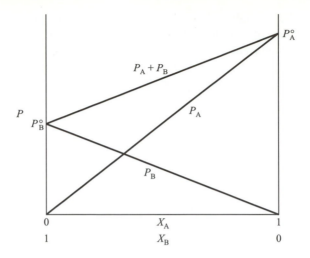

Figure 7.4 A plot of the vapour pressures, at some arbitrary temperature, against the composition of the liquid, for the hypothetical system, *A–B*, deemed to obey Raoult's Law.

$$
\left.
\begin{aligned}
y_A &= \frac{P_A}{P_A + P_B} = \frac{X_A P_A^\circ}{X_A P_A^\circ + X_B P_B^\circ} \\
y_B &= \frac{P_B}{P_A + P_B} = \frac{X_B P_B^\circ}{X_A P_A^\circ + X_B P_B^\circ}
\end{aligned}
\right\}
\tag{7.16}
$$

It can readily be seen from Equation (7.16) that, if $P_A^\circ > P_B^\circ$, then at $X_A = X_B = 0.5$, $y_A > 0.5$. In fact, if $P_A^\circ > P_B^\circ$, then y_A is always greater than X_A (except, of course, at $X_A = 0$ and $X_A = 1!$). Consequently, the plot of y_A against P is not a straight line, but a curve, concave upwards. This means that, as an inevitable consequence of Raoult's Law, the vapour phase is always richer than is the liquid phase in the more volatile of the two components.

—— **EXAMPLE 7.2**——————————————————————————
At 298 K, the densities of ethanol and water are 0.785 and 0.997 g cm^{-3}, and the respective vapour pressures are 57.2 and 23.7 torr. A mixture of these two liquids is prepared at 298 K by mixing equal volumes of the two components.

(i) What is the mole fraction of ethanol in this mixture?

(ii) Assuming Raoult's Law is obeyed, what would be the total vapour pressure of this mixture?

(iii) Assuming Raoult's Law is obeyed, what would be the composition of the vapour in equilibrium with this liquid mixture?

— SOLUTION

We need to evaluate the number of moles of each component in some arbitrary volume, let us say in 100 cm^3. To do this, we divide the mass in g by the molar mass in g mol^{-1}.

$$\begin{aligned}
\text{Amount of ethanol} \quad &= 100 \text{ cm}^3 \\
&\equiv 100 \text{ cm}^3 \times 0.785 \text{ g cm}^{-3} \\
&\equiv \frac{(100 \times 0.785) \text{ g}}{46.07 \text{ g mol}^{-1}} = 1.704 \text{ mol}
\end{aligned} \qquad (7.17)$$

$$\begin{aligned}
\text{Amount of water} \quad &= 100 \text{ cm}^3 \\
&\equiv 100 \text{ cm}^3 \times 0.997 \text{ g cm}^{-3} \\
&\equiv \frac{(100 \times 0.997) \text{ g}}{18.02 \text{ g mol}^{-1}} = 5.533 \text{ mol}
\end{aligned} \qquad (7.18)$$

Thus the mole fractions of the liquid mixture are:

$$\begin{aligned}
X_{\text{ethanol}} &= \frac{1.704}{1.704 + 5.533} \\
&= 0.235
\end{aligned} \qquad (7.19)$$

$$\begin{aligned}
X_{\text{water}} &= \frac{5.533}{1.704 + 5.533} \\
&= 0.765
\end{aligned} \qquad (7.20)$$

These constitute the answers to part (i).

If Raoult's Law is obeyed, then the total vapour pressure P is given by the sum of $X_A P_A^o$ and $X_B P_B^o$. In the present context, this yields:

$$\begin{aligned}
P &= (0.235 \times 57.2 \text{ torr}) + (0.765 \times 23.7 \text{ torr}) \\
&= (13.4 + 18.1) \text{ torr} \\
&= 31.5 \text{ torr}
\end{aligned} \qquad (7.21)$$

We note that this figure lies, as it must, between those of the two pure components, and that it answers part (ii).

The molar composition of the vapour is obtained from the partial pressures of the two components:

$$\begin{aligned}
y_{\text{ethanol}} &= \frac{13.4}{13.4 + 18.1} \\
&= 0.425
\end{aligned} \qquad (7.22)$$

$$\begin{aligned}
y_{\text{water}} &= \frac{18.1}{13.4 + 18.1} \\
&= 0.575
\end{aligned} \qquad (7.23)$$

119

Clearly, the mole fraction of ethanol in the vapour is well in excess of its mole fraction in the liquid, as one must expect since ethanol is the more volatile of the two components.

Solutions which obey Raoult's Law are labelled ideal solutions, and are in the minority. This behaviour occurs when the intermolecular forces in the two liquid components are very similar, as happens, for example, with toluene and p-xylene. In most cases, and this includes the system ethanol–water, deviations are found from the behaviour prescribed in Equations (7.15). Two alternative patterns of deviation are experienced. In one group, the actual values of P_A and P_B are (except at $X_A = 0$ and $X_A = 1$) consistently less than prescribed above, so this is classified as negative deviation from Raoult's Law. Where this occurs, the curve of the total vapour pressure, P, against mole fraction usually exhibits a minimum. The other group show the opposite behaviour, known as positive deviation, where the values of P_A and P_B are greater than is implied by Equations (7.15), and P usually exhibits a maximum.

These deviations tend to arise where the attractive forces between molecules A and B are substantially different from the mean of those between A and A and those between B and B. Whether the A–B forces are weaker or stronger than the mean of those within the pure components, substantial deviations arise only where components A and B are molecules of essentially different types.

7.6 Fractional distillation of a liquid mixture

On the basis of the behaviour summarised by Raoult's Law, a liquid mixture should be capable of separation into its pure components by the process of fractional distillation. The purpose of this section is to explain how this can be achieved.

Figure 7.5 is essentially a repetition of Figure 7.4, in that it shows the variation of the total pressure, P, as a function of the molar composition of the liquid. Another line, in grey, has been added showing the interrelationship between P and the composition of the vapour phase. (We are well used to graphs in which different curves show the change, against a single variable as abscissa, of various functions plotted as the ordinate. The inherent difficulty about Figure 7.5 is that we have a single parameter, P, as ordinate but two alternative functions as abscissae.) For any value of P other than at the extreme limits, the vapour is richer than is the liquid in the more volatile of the two components, which means that, in this instance, $y_A > X_A$.

As we saw in section 7.3, the vapour pressure of a liquid increases with increasing temperature, and when the vapour pressure has attained that of the prevailing atmosphere, the liquid will boil. What we now want to do is to look, not at the variation of the vapour pressure at constant temperature,

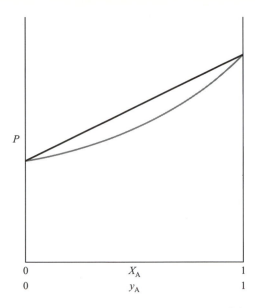

Figure 7.5   Plots of (black line) the total vapour pressure, P, of the A–B system against the composition of the liquid phase and of (grey line) P against the composition of the vapour in equilibrium with this liquid, assuming Raoult's Law.

but rather at the variation of the boiling point at constant atmospheric pressure. For a volatile liquid, the vapour pressure is high so the boiling point will be low, and vice versa. So the curve of the boiling point against composition is, to some extent, a reversal of Figure 7.5, and this is shown in Figure 7.6. Once again, we have two lines: the black one is of boiling point against the mole fraction of the liquid and the grey one is of the boiling point against the mole fraction of the vapour. Both of these lines are curved. As in Figure 7.5, at the same temperature the value of y_A is greater than X_A.

Suppose we have a binary liquid mixture whose X_A value corresponds to the dotted vertical line in Figure 7.6. If this is put into the flask of a distillation apparatus, and sufficient heating is applied, the liquid will boil. The requisite temperature will be where the dotted line intersects the black curve. The composition of the vapour in equilibrium with the liquid at this temperature will be given by the value of the grey line at the same temperature. So, referring to the sketch of the distillation apparatus in Figure 7.7, rising from the surface of the liquid at a there will be vapour of composition y_1.

As the vapour rises up the unheated distillation column, it will condense on the surface at some point, b, inevitably producing a liquid of that same composition. On Figure 7.6, the dotted vertical at y_1 indicates the position now achieved. Since heating is continually being applied to the flask, a stage will be reached where vapour progresses above the point b. This vapour has to be in

121

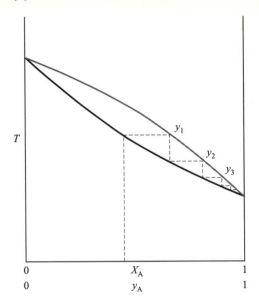

Figure 7.6 Plots of the boiling point of the liquid *A–B* system against (black line) the composition of the liquid phase and (grey line) the composition of the vapour phase in equilibrium with it. The dotted lines are explained in the text.

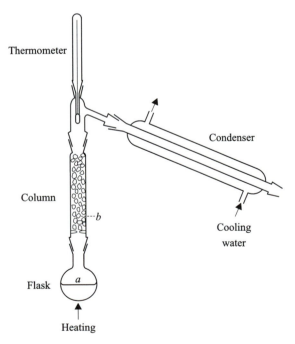

Figure 7.7 A schematic diagram of an apparatus for the fractional distillation of a liquid.

equilibrium with the liquid condensed at b, so on the basis of Figure 7.6, its composition will be given by y_2.

The argument of the previous paragraph can be repeated and leads to the attainment, farther up the column, of compositions y_3, then y_4, and so on. The end result must be that the vapour going past the thermometer bulb at the top of the distillation column has $y_A = 1$, that is, it is pure component A. So, over the initial period, the distillate will be pure A, and, in consequence, the residue becomes progressively richer in component B, and eventually it becomes pure B.

However, not all pairs of liquids obey Raoult's Law and, unless the deviation is very slight, it is almost inevitable that P will show a turning value. If P displays a maximum, then the corresponding boiling point will be a minimum, and vice versa. In consequence, one is not able to obtain both pure components by fractional distillation.

As an illustration, let us consider the system ethanol–water, which shows positive deviations from Raoult's Law. Ethanol is the more volatile of these components, but the distillate has the composition of the minimum in the boiling point curve, which is that of the maximum in the vapour pressure, with a mole fraction of ethanol equal to 0.904. This is called the azeotrope, because at this point the liquid and the vapour in equilibrium with it have the same composition, so that there is no change in the composition of such a liquid as it boils. As one distills off the low-boiling azeotrope, the residue becomes progressively richer in water.

7.7 The solubility of gases in liquids

When a gas is in contact with a liquid and agitation is applied to achieve equilibrium, dissolution of the gas in the liquid occurs to an extent which is proportional to the partial pressure of the gas, as is illustrated in Figure 7.8. As a consequence of the same partial pressures in the gas phase, different gases achieve different concentrations in solution.

Several different systems have been used by scientists to report the solubility of a particular gas in a solvent. One of these is Henry's Law, which may be written,

$$K = \frac{P}{X} \tag{7.24}$$

where P denotes the partial pressure of the gas, and X its mole fraction in the solution. It follows that K, called the Henry's Law constant, has the units of pressure and its value is inversely proportional to the mole fraction achieved at unit pressure. Usually, but not invariably, the value of K increases with increasing temperature.

Before engaging in any calculations about the equilibrium concentrations of dissolved gases, it should be stressed that dissolution equilibria are not readily

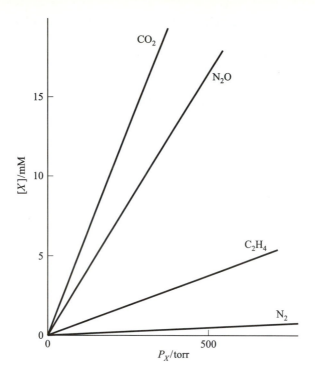

Figure 7.8 Plots of the equilibrium concentrations of various gases achieved in aqueous solution at 293 K, as a function of the partial pressure of the gas.

achieved. Chiefly, this is because the diffusion of a solute occurs so slowly that considerable agitation is needed to help achieve a uniform concentration. Likewise, if the concentration of a gas in solution is higher than the value that would be in equilibrium with the partial pressure of the gas above it, agitation assists in rapidly achieving equilibrium by having gas leave the solution. This fact is well attested by the actions of successful Grand Prix drivers just after receiving their awards on the winners' rostrum.

— **EXAMPLE 7.3** —————————————————————————
At 293 K, the Henry's Law constant for O_2 in water is 2.95×10^7 torr. What partial pressure is required so that the equilibrium concentration of O_2 in water at 293 K will be 1×10^{-3} mol dm^{-3} ?

— **SOLUTION** ——————————————————————————
First, we want to express the desired concentration of O_2 as a mole fraction. Consider 1 dm^3 of solution, containing 10^{-3} moles of O_2 and 1000/18.0 moles of water.

124

$$X_{O_2} = \frac{10^{-3}}{55.55} = 1.80 \times 10^{-5} \tag{7.25}$$

Rearranging Equation (7.24), and substituting, we have:

$$P = K.X$$

$$= 2.95 \times 10^7 \text{ torr} \times 1.80 \times 10^{-5} \tag{7.26}$$

$$= 531 \text{ torr}$$

The solubilities of monatomic (e.g. Ar) and homonuclear diatomic (e.g. H_2, N_2, O_2) gases in water are all fairly low. Heteronuclear diatomics like CO or NO, and polyatomic molecules like N_2O or CO_2 tend to be rather more soluble. In a few cases, such as HCl, the interaction involved is not just one of strong intermolecular forces but in effect of a chemical reaction, so these gases show much higher solubilities.

If one agitates a system containing a liquid and a mixture of gases, then the concentrations in solution of each component will tend towards the equilibrium values in accordance with the partial pressures. Bubbling nitrogen or argon through a solution will achieve the equilibrium concentration for 760 torr of that gas, and also serves to remove oxygen from solution, since the partial pressure of O_2 in contact with the solution has been reduced to zero.

7.8 Colligative properties of solutions

There are three well-established properties of dilute solutions which have certain aspects in common and are frequently known by the above umbrella title, derived from the Latin word meaning 'gathered together'. These are the depression of the freezing point of a solution, the elevation of the boiling point of a solution and the osmotic pressure of a solution. The common feature is that, in each case, the magnitude is a function only of the molar concentration of the solute, regardless of its identity.

The easiest of the three to explain is perhaps the elevation of the boiling point. We assume that the solute B is involatile and that the solution obeys Raoult's Law. So whereas the vapour pressure of the pure solvent is P_A^o, that of the solution, in which the solvent has a mole fraction of $(1 - X_B)$, is given by $(1 - X_B) P_A^o$. Because of this depression of the vapour pressure, a higher temperature will be required for the solution to achieve a vapour pressure of 760 torr. The magnitude of this increase in the boiling point, ΔT, will be proportional to X_B and, bearing in mind the Clausius–Clapeyron equation, it will be inversely proportional to the latent heat of vaporisation of the solvent. Empirically, we have,

$$\Delta T = K_b \cdot m_B \tag{7.27}$$

where m_B denotes the molality of the solute (i.e. the number of moles per kg of solvent) and K_b, called the molal ebullioscopic constant, is a function of the solvent.

Similarly, for the depression of the freezing point, we have the analogous empirical expression,

$$-\Delta T = K_f \cdot m_B \tag{7.28}$$

where K_f denotes the molal cryoscopic constant, which is inversely proportional to the latent heat of fusion of the solvent.

In general, latent heats of vaporisation are quite large, leading to low values for K_b for most solvents. Latent heats of fusion are mostly smaller, leading to rather larger values for K_f. So if we confine ourselves to dilute solutions, larger changes are usually experienced in the freezing point than in the boiling point. For this reason it is the depression of freezing point which is the more useful in finding the effect in solution of a certain mass of an unknown substance, as a means of establishing the molar mass.

Whereas the K_f value for benzene is 5.1 K kg mol^{-1}, that for camphor is 38 K kg mol^{-1}. This disparity arises because the latent heat of fusion of camphor is anomalously low, for a reason which is well understood but for which we have not covered the background. Additionally, camphor melts at the readily accessible temperature of 178°C, whereas benzene freezes at 5.4°C. For these reasons, the use of the freezing point depression to establish the molar mass of an organic substance is usually achieved using camphor, by what is known as Rast's method.

The K_f value for water is 1.86 K kg mol^{-1}, so very large amounts of solute are needed to depress the freezing point of water beyond the limits attainable in a cold winter. In fact, the amount required is so large that the contents of a car radiator could hardly be classed as a dilute solution and Equation (7.28) is not strictly applicable.

— EXAMPLE 7.4 —
When 1.5 g of an unknown, involatile solid was added to 100 g of benzene, the freezing point of the latter was depressed by 0.42 K. Evaluate the molar mass of the unknown substance.

— SOLUTION —
1.5 g in 100 g of solvent $\equiv$ 15 g in 1 kg.
So the molality of the solution is given by,

$$m_B = \frac{15}{M}$$

where M is the molar mass of the solute. Substituting into Equation (7.28), we have:

$$0.42 = 5.1 \times \frac{15}{M} \tag{7.29}$$

$$\therefore M = \frac{5.1 \text{ K kg mol}^{-1} \times 15 \text{ g kg}^{-1}}{0.42 \text{ K}} = 182 \text{ g mol}^{-1} \tag{7.30}$$

7.9 The osmotic pressure

The phenomenon of osmosis was initially investigated using a piece of pig's bladder, whose role was later recognised as that of a semipermeable membrane, porous to the solvent but impermeable to the solutes. In later work, synthetic membranes were usually used, so that the advance of science would not seem so inimical to the interests of pigs.

The basic phenomenon is shown diagrammatically in Figure 7.9. When there is a solution on one side of a semipermeable membrane and the pure solvent on the other, transference of the solvent occurs so that a difference in hydrostatic pressure on the two sides of the membrane is achieved. At equilibrium, this pressure difference will be equal and opposite to the osmotic pressure, π, of the solution. In the present case, the hydrostatic pressure can

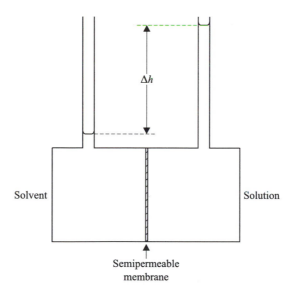

Figure 7.9 A schematic diagram of a system to illustrate the osmotic pressure, with a solution to one side and the solvent only to the other side of the semipermeable membrane. The vertical difference in the levels of these two liquids is denoted by Δh.

be measured as the product, $\Delta h.\rho.g$, where Δh is the difference in the liquid levels, ρ is the density of the solvent and g is the acceleration due to gravity.

Van't Hoff showed that the osmotic pressure is proportional to the concentration, c, of the solute and to the Kelvin temperature. In fact, it fits the equation,

$$\pi = cRT \tag{7.31}$$

where R is the gas constant and all quantities are expressed in mutually consistent units. If the concentration, c, is replaced by the number of moles, n, divided by the volume, V, we obtain,

$$\pi V = nRT \tag{7.31a}$$

which closely parallels equation (5.1) for the behaviour of a perfect gas.

The magnitude of the osmotic pressure of a moderately dilute solution can be appreciable. Suppose 1.5 g of sucrose (M = 342 g mol^{-1}) is added to 100 cm^3 of water at 60°C. The osmotic pressure can readily be evaluated from Equation (7.31), using SI units for each parameter.

$$\pi = \frac{1.5}{342} \times 10^4 \text{ mol m}^{-3} \times 8.314 \text{ J K}^{-1} \text{ mol}^{-1} \times 333 \text{ K}$$
$$= 1.21 \times 10^5 \text{ Pa} \tag{7.32}$$
$$\equiv 1.20 \text{ atm}$$

The factor of 10^4 in the first line of the above calculation arises because the SI unit of volume is 1 m^3, and 1.5 g per 100 cm^3 means 10^4 times this per m^3. So c in SI units is multiplied by R and by T, also in SI units, to give the pressure in SI units, namely N m^{-2} or Pa.

Van't Hoff's studies of the osmotic pressure also embraced the behaviour of electrolytes, which are discussed in Chapter 9. Here, he found that an amendment to Equation (7.31) was necessary and he introduced the factor i to the equation, so that, for electrolyte solutions, it reads:

$$\pi = icRT \tag{7.33}$$

(Alternatively, one may take the view that the Van't Hoff i factor is always present, but for non-electrolytes it is always unity.) For electrolytes, the i factor is greater than one, usually but not invariably an integer, such as 2 or 3. It is readily rationalised on the basis that it represents the number of entities present in solution for each 'molecule' that is dissolved. For sucrose, this is one, but for NaCl, which yields the ions Na$^+$ and Cl$^-$ in solution, i is equal to two.

As the origins of our knowledge of the osmotic pressure indicate, this is a very significant parameter within biological systems. For patients in various conditions, there is a need to be able to put fluid into the bloodstream via a vein, to counteract dehydration. Desirably, the solution to be injected should have an osmotic pressure to match that of blood. If water were injected, the red cells in the blood would gain water by osmosis, probably until they burst. Sterile packs labelled 'saline drip' contain 9 g of NaCl per dm^3, and we may evaluate the osmotic pressure at body temperature (37°C) using Equation (7.33).

$$\pi = 2 \times \frac{9.0}{58.4} \times 10^3 \text{ mol m}^{-3} \times 8.314 \text{ J K}^{-1} \text{ mol}^{-1} \times 310 \text{ K}$$
$$= 7.94 \times 10^5 \text{ Pa} \tag{7.34}$$
$$\equiv 7.8 \text{ atm}$$

7.10 Solute partition between immiscible liquids

Suppose we have two immiscible liquids and add to the system a solute which dissolves in both liquids. How will it be distributed between the two phases? The answer is that, at equilibrium (which, again, will be achieved only after agitating the system) the ratio of the concentrations in the two phases will be a constant. This ratio is known as the distribution coefficient, K_D, and has a value peculiar to the solute and the pair of solvents.

If the solute is a gas, then the constancy of the ratio of the concentrations in the two phases may be seen as an inevitable consequence of Henry's Law. In each phase, the concentration of the gas will be proportional to the partial pressure, so the two concentrations must be in a constant ratio.

The fundamental reason for the dissolution of a solute in a solvent is, of course, that in this way the system achieves a lower Gibbs energy. As we shall see in Chapter 8, the Gibbs energy per mole of a solute increases with the concentration of the solution. On the other hand, the Gibbs energy per mole of a solid is invariant. So when a solid solute is added to a solvent, it will dissolve only until the stage where the Gibbs energy per mole of the solute in solution is equal to that of the solid. The solution is then saturated and this specifies the solubility of this solute.

For the system of two immiscible liquid phases, the above paragraph is applicable to the dissolution of the solute in both phases. If some solid solute is present, then there must be equilibrium between the solid and the dissolved solute in each phase. This means that the distribution coefficient is equal to the ratio of the solubilities of the solute in the two phases, so that K_D is in no sense an arbitrary parameter.

This partition law is the basis for solvent extraction procedures. Suppose we have a mixture of substances dissolved in water, of which there is interest only in one. If we can identify a solvent, immiscible with water, in which this substance and only it is highly soluble, then it is a simple matter to add this solvent to the aqueous solution, shake them up to achieve equilibrium and separate off the non-aqueous layer. This will be rich in the desired substance, to an extent which will depend on its relative solubilities in the two solvents.

Suggested reading

FRANKS, F., 1983, *Water*, London: The Royal Society of Chemistry.
HILDEBRAND, J. L., PRAUSNITZ, J. M. and SCOTT, R. L., 1970, *Regular and Related Solutions*, New York: Van Nostrand Reinhold.
CHANG, R. and SKINNER, J. F., 1990, Ice under pressure, *Journal of Chemical Education*, **67**, 789–90.
BONICELLI, M.G., di GIACOMO, F., CARDINALE, M. E., and CARELLI, I., 1984, A new determination of the approximate phase change formula, $\Delta T = K.m.$, *Journal of Chemical Education*, **61**, 423–24.
LOGAN, S. R., 1998, The behavior of a pair of partially miscible liquids, *Journal of Chemical Education*, **75**, 339–42.

Problems

7.1 The vapour pressure of (liquid) lead varies with temperature as follows:

$t/°C$	973	1162	1421
$P/torr$	1.0	10	100

Evaluate the molar enthalpy of vaporisation and the boiling point of lead.

7.2 The vapour pressure of tetrachloromethane (CCl_4) varies with temperature as follows:

$t/°C$	P/atm	$t/°C$	P/atm
0	0.043	40	0.284
10	0.074	50	0.417
20	0.120	60	0.593
30	0.188	70	0.818

Use these data to evaluate (a) the molar latent heat of vaporisation of CCl_4, (b) the boiling point of CCl_4, and (c) the vapour pressure of CCl_4 at its freezing point, $-23°C$.

7.3 At 323 K, a mixture of benzene (V.P. = 268 torr) and 1,2-dichloroethane (V.P. = 236 torr) behaves ideally. What is the total vapour pressure of a liquid mixture consisting of (a) equimolar quantities, and (b) equal weights, of the two components? In each case, evaluate the molar composition of the vapour. (C_6H_6 = 78.1; $C_2H_4Cl_2$ = 99.0)

7.4 Chlorobenzene and bromobenzene form an ideal solution. At 400 K, their respective vapour pressures are 115.0 and 60.4 kPa. Calculate the composition of the binary mixture which boils at precisely 400 K, and the molar composition of the vapour in equilibrium with it.

7.5 At 40°C, the vapour pressures of pure chloroform ($CHCl_3$) and pure acetone (CH_3COCH_3) are respectively 360 and 410 torr. At this temperature, a solution containing 2 mol of chloroform and 3 mol of acetone has a total vapour pressure of 350 torr.

Evaluate the total vapour pressure to be expected at this temperature if Raoult's Law were obeyed. Describe the type of deviation shown by this solution and suggest reasons for it.

7.6 The Henry's Law constant for N_2O in water at 293 K is quoted as 0.155×10^7 torr. If I bubble N_2O through a test tube of water at this temperature, what molar concentration may I expect to achieve? Is the answer to the previous part more or less than the molar concentration of gaseous N_2O at 1 atm pressure and 293 K?

7.7 A diver is working under water at a depth of 250 m, and is breathing a mixture of helium and oxygen, suitably compressed to accommodate him. What should be the mole fraction of oxygen in the mixture so that the diver has the same concentration of oxygen in his blood as he would have if he were breathing fresh air at the water's surface. Oxygen comprises 20.9 mole % of the atmosphere.

7.8 At 293 K, the vapour pressure of CS_2 is 85.40 torr. When 2.0 g of sulfur is dissolved in 100.0 g of CS_2, the vapour pressure at 293 K is 84.90 torr. Evaluate the molar mass of the solute. (Molar mass of CS_2 = 76.12 g mol^{-1})

7.9 As a consequence of a lengthy series of laboratory operations, you have isolated a new antibiotic, whose molar mass appears to be around 10 000 g mol^{-1}. It is desired to check this by a colligative method, but very little of the substance can be spared, just enough for a 1% by weight aqueous solution. Calculate the anticipated

(a) depression of the freezing point,
(b) elevation of the boiling point,
(c) decrease of the vapour pressure at 298 K, and
(d) osmotic pressure at 298 K

For water, the molal ebullioscopic constant is 0.51 K kg mol^{-1}, the molal cryoscopic constant is 1.86 K kg mol^{-1} and the vapour pressure at 298 K is 23.76 torr.

7.10 At 298 K, a solution prepared by dissolving 71.5 mg of a non-electrolyte in 100 cm^3 of water has an osmotic pressure of 88.0 torr. Evaluate the molecular weight of the non-electrolyte. (1 atm $\equiv$ 760 torr $\equiv$ 1.013 $\times 10^5$ Pa)

17.11 A solution containing 122 g of benzoic acid, $C_6H_5CO_2H$, in 1000 g of benzene boils at 81.5°C, whereas the boiling point of benzene is 80.1°C. Find the apparent molar mass of benzoic acid in this solution and suggest a reason for the discrepancy between it and the molecular formula given above. For benzene, the boiling point elevation constant, K_b = 2.53 K kg mol^{-1}.

8

Chemical equilibrium

When the principle of the minimising of the Gibbs energy is applied to chemical processes, for reactions within a single homogeneous phase (such as the gas phase or a solution), this leads to the concept of a finite equilibrium constant, invariably denoted by K and varying only with the temperature. This variation permits the standard enthalpy change and the standard entropy change for the reaction to be evaluated. Heterogeneous reactions involving two solid phases are shown to behave differently and to have a sharp transition temperature, which is the only condition at which the solid reactant and the solid product can coexist.

8.1 Equilibrium between two solids

Sometimes a substance can crystallise in either of two forms. For example, sulfur may adopt either the rhombic (called α) or the monoclinic (β) crystal structure. The question is, which will be adopted? The answer, very simply, is that at constant temperature and pressure, the more stable form is that with the lower value of the Gibbs energy, G. This applies in general to chemical equilibrium between solids.

To illustrate this point, we may refer to Figure 8.1, which shows how G per mole varies with temperature for these two forms of sulfur. Below 95.5°C, the α form has the lower value of the molar Gibbs energy, whereas above this temperature, it is the β form which offers the lower value of G. So α-sulfur is stable below 95.5°C and β-sulfur is stable above this temperature, which is the transition point.

On the more general question, one qualification should be added. The only processes that will occur will be those for which the Gibbs energy decreases. While this is a necessary condition, its fulfilment does not guarantee that, regardless of the temperature, the process will occur at a perceptible rate. In

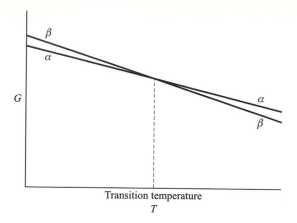

Figure 8.1 Illustration of the variation with the temperature, T, of the molar Gibbs energy, G, for two solids which may undergo interconversion at a transition temperature.

a way, we are fortunate that this is so. Otherwise, a delicious meal would react on the plate with the oxygen of the air, leaving only carbon dioxide, water and oxides of nitrogen!

8.2 Reactions between gases

The Gibbs energy of one mole of a gaseous substance depends on the pressure of the gas. The relevant equation is easily deduced from Equation (6.17a). If we keep the temperature constant, the last term vanishes, leaving $dG = V.dP$. Substituting for P from the equation for one mole of a perfect gas, we obtain:

$$dG = RT\frac{dP}{P} \tag{8.1}$$

On integration, this leads to:

$$G = RT\ln P + \text{const} \tag{8.2}$$

So, for one mole of a perfect gas, the Gibbs energy increases with increasing pressure in accordance with the first term on the right. Since the logarithm of one is zero, this term vanishes when P has unit value. It is convenient to regard the Gibbs energy at unit pressure as the standard value, written with a superscript plimsoll, '$G^{\ominus}$' and the standard pressure is taken as 1 bar $= 10^5$ Pa $= 750$ torr. So we have:

$$G = G^{\ominus} + RT\ln P \tag{8.2a}$$

This implications of this variation of the molar Gibbs energy with the pressure are shown in Figure 8.2.

134

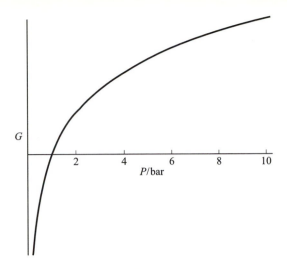

Figure 8.2 Plot of equation (8.2a), showing how the molar Gibbs energy, G, varies with the pressure, P, of a perfect gas.

Let us consider the gas phase reaction process,

$$2CO + O_2 \quad = \quad 2CO_2 \tag{8.3}$$

from the aspect of the position of equilibrium in a system of this nature. Firstly, we can say that this will be achieved when the Gibbs energy of the system is minimised. It follows that, at equilibrium, the Gibbs energy of the molar quantities on the left-hand side of the stoichiometric equation are equal to that of the right-hand side:

$$2G_{CO} + G_{O_2} = 2G_{CO_2} \tag{8.4}$$

Substituting into Equation (8.4) from Equation (8.2a) and its analogues, we have:

$$2G^{\ominus}_{CO} + 2RT \ln P_{CO} + G^{\ominus}_{O_2} + RT \ln P_{O_2} = 2G^{\ominus}_{CO_2} + 2RT \ln P_{CO_2} \tag{8.5}$$

Rearranging, and making use of the properties of logarithms expounded in Appendix Two, we have:

$$2G^{\ominus}_{CO_2} - 2G^{\ominus}_{CO} - 2G^{\ominus}_{O_2} = -RT \ln \left(\frac{P^2_{CO_2}}{P^2_{CO} P_{O_2}} \right) \tag{8.6}$$

The left-hand side of this equation is a constant and so must be the right-hand side. This means that the quotient in the brackets on the right-hand side is a constant. It is called the equilibrium constant, K. In this case, since every factor within it is a pressure, it is written K_p, called the equilibrium constant in terms of pressures.

In Equation (8.6), the left-hand side is the standard Gibbs energy of the products of the stoichiometric equation, (8.3), less that of the reactants. For short we may call it $\Delta G^{\ominus}$, the standard Gibbs energy change, and write the equation for equilibrium as:

$$\Delta G^{\ominus} = -RT \ln K_p \qquad (8.6a)$$

For a reaction such as Equation (8.3), $\Delta G^{\ominus}$ must have a finite value, positive or negative. If positive, then K_p must be less than one, which means that if the reactant pressures are initially around unit pressure, then only a small extent of reaction is possible on thermodynamic grounds. If $\Delta G^{\ominus}$ is negative, then K_p must exceed one and at equilibrium the reaction must lie over on the product side. In either case, the position of equilibrium will correspond to the attainment of some specified value for K_p, and the extent to which reaction will have occurred will be finite. Unlike the reaction involving two solids, gas phase reactions cannot go to completion.

From tabulated data, $\Delta G^{\ominus}$ at 298 K for Reaction (8.3) can be reckoned as -513.9 kJ mol^{-1}, which leads to a value of 1.1×10^{90} for K_p at that temperature. In a normal atmosphere with a partial pressure of O_2 equal to 0.21 bar, then at equilibrium we can calculate that:

$$\frac{P_{CO_2}}{P_{CO}} = 4.8 \times 10^{44} \qquad (8.7)$$

This would seem to suggest that, in the atmosphere at ambient temperatures, the deadly carbon monoxide gas will be converted, almost quantitatively, into the relatively safe carbon dioxide. Unhappily for those with faulty gas appliances, the reservations in the last paragraph of the previous section still apply. At ambient temperatures, Reaction (8.3) does not proceed at a perceptible rate towards equilibrium. Any CO which escapes from the gas heater into the room will remain as CO. At flame temperatures, the value of $\Delta G^{\ominus}$ for Reaction (8.3), although not identical to that quoted above, is only marginally different, so K_p remains very large. If an abundance of oxygen is present in the flame then, when equilibrium is achieved at flame temperature, very little carbon monoxide will be left in the exhaust gases.

8.3 Reactions in solution

In solution, the Gibbs energy per mole of a solute rises with concentration just as does that of a gas. For ideal solutions, G is assumed to vary with the concentration in a manner analogous to that of a perfect gas. So for solute A, we have,

$$G_A = G_A^{\ominus} + RT \ln [A] \qquad (8.8)$$

where $[A]$ denotes the concentration of A in the solution and $G_A^{\ominus}$ the standard Gibbs energy per mole at unit concentration.

For a reaction in solution, the conditions governing equilibrium are very similar to those operative in the gas phase. Considering the empirical reaction,

$$A + B = C \tag{8.9}$$

for each substance an equation analogous to Equation (8.8) is applicable. This leads to the relation,

$$G_C - G_A - G_B = G_C^{\ominus} \, p1 + RT \, \ln [C] - G_A^{\ominus} - RT \, \ln [A] - G_B^{\ominus} - RT \, \ln [B] \tag{8.10}$$

where $[C]$ denotes the actual concentration of species C and G_C the molar Gibbs energy of this species at this concentration.

This may more succinctly be written,

$$\Delta G = \Delta G^{\ominus} + RT \, \ln \frac{[C]}{[A][B]} \tag{8.10a}$$

where ΔG is the actual molar Gibbs energy change for the reaction and ΔG^{θ} is the standard Gibbs energy change, that is, the change in Gibbs energy when the molar quantities of the reactants in Equation (8.8) are converted into the specified molar quantity of the product, with each species at its standard state, that is, at unit concentration.

Where Reaction (8.8) is at thermodynamic equilibrium, we may make two changes to Equation (8.10a). Firstly, we may equate ΔG to zero; secondly, the concentration quotient in the last term will have become the equilibrium constant in terms of concentrations, K_c. This gives:

$$\Delta G^{\ominus} = -RT \ln K_c \tag{8.10b}$$

The value of $\Delta G^{\ominus}$, whether it be positive or negative, will be finite, and so must K_c. At equilibrium, the species A, B and C must all be present in the solution.

Let us consider the solution equilibrium,

$$I^- + I_2 = I_3^- \tag{8.11}$$

for which, in aqueous solution at 298 K, the equilibrium constant is 7.4×10^2. Suppose we add to the solvent amounts of KI and of I_2 which would, in the absence of complex formation, give respective concentrations of a and b. In the event, on account of Reaction (8.11), these are not attained. Let us assume the concentration of the I_3^- formed is x. We have:

$$K_c = \frac{[I_3^-]}{[I^-][I_2]}$$
$$= \frac{x}{(a - x)(b - x)} = 740 \tag{8.12}$$

If b is much greater than a, this may be simplified in that x, which is less than b, is negligible in relation to a, so for $(a - x)$ we write a.

This leads to:

$$x = \frac{740ab}{1 + 740a} \tag{8.13}$$

Putting $a = 10^{-2}$ mol dm^{-3} and $b = 5 \times 10^{-4}$ mol dm^{-3}, this leads to $x = 4.4 \times 10^{-4}$ mol dm^{-3}. This means that, for this concentration of KI, 88% of the 'iodine' exists as the I_3^- complex and only 12% as the I_2 molecule. A higher KI concentration would lead to a greater degree of complexation.

Equilibrium (8.11) was formerly a familiar one in domestic first aid cabinets, under the name 'tincture of iodine'. This was widely used as an antiseptic on abrasions of the skin. The addition of the I^- ion not only allowed far more 'iodine' to be held in solution, but retarded its rate of oxidation. Such solutions were often kept for years. The distinctive yellowish-brown colour of the remedy is of course due to the I_3^- ion.

8.4 The significance of the phase of the reacting species

The designated phases of the reacting species considered in the three preceding sections do not exhaust all possible scenarios, but they demonstrate that a special situation exists when one has a solid species on each side of the stoichiometric equation. The serious student may wonder why this is so, and we will offer two alternative lines of explanation, which are basically equivalent.

With a solid species, the Gibbs energy per mole shows no dependence on the quantity present. If the Gibbs energy per mole of one phase is greater than that of the other, interconversion is possible on thermodynamic grounds and this will remain the case until all of that phase is used up. On the other hand, if the reactants are in a homogeneous phase, then the Gibbs energy per mole varies with the concentration, as demonstrated in Figure 8.2. As this species is used up, its Gibbs energy per mole falls and at some stage further reaction becomes impossible.

Another standpoint is to consider the Gibbs energy of the total system of reactants and products. Its tendency to seek a minimum will determine the extent of reaction at equilibrium. Where there are two solids involved, these do not intermix, so that the total Gibbs energy is the sum of two contributions which are both proportional to the amount of that solid present. So the plot of G against the extent of reaction will be a straight line and the system will tend towards that end of the line where G is the lower. But where the reaction is homogeneous, the plot of G against the extent of reaction is always a curve, concave upwards, so that the minimum Gibbs energy towards which the system tends will have some finite extent of conversion. The reason for this curvature is that in the homogeneous system there is a significant and negative contribution to the Gibbs energy from the mixing of the reactants and the products. In the heterogeneous case, no mixing occurs so this contribution is absent. The contrast is highlighted in Figure 8.3.

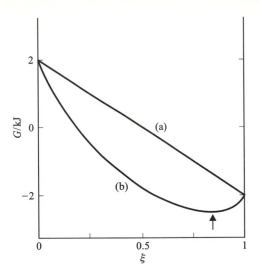

Figure 8.3 Plots of the molar Gibbs energy, G, against the degree of interconversion, ξ, between two species of identical molecular formulae, for an assumed value of the standard Gibbs energy change, $\Delta G^{\ominus}$: (a) a solid–solid transition; (b) an equilibrium in a homogeneous phase.

8.5 The effect of temperature on chemical equilibria

The earlier discussions of systems approaching chemical equilibrium were conducted in terms of constant pressure and constant temperature. It is pertinent to inquire how the value of this constant temperature affects the outcome.

For reactions in a homogeneous phase, the condition of equilibrium is readily summarised by the equation,

$$\Delta G^{\ominus} = -RT \ln K \tag{8.14}$$

where K denotes the appropriate equilibrium constant. (See Equations 8.6a and 8.10.) Remembering the definition of the Gibbs energy (Equation 6.16), the left-hand side of this relation may be expressed in terms of the standard changes in the enthalpy, H, and the entropy, S:

$$\Delta H^{\ominus} - T.\Delta S^{\ominus} = -RT \ln K \tag{8.15}$$

Dividing across by RT, this leads to:

$$\ln K = -\frac{\Delta H^{\ominus}}{RT} + \frac{\Delta S^{\ominus}}{R} \tag{8.16}$$

Strictly speaking, $\Delta H^{\ominus}$ and $\Delta S^{\ominus}$ are not independent of temperature but, as we saw in p. 101, the standard enthalpy change varies only slightly as the temperature is changed. The entropies of most substances increase as the temperature is increased, but since this will apply both for the reactants and the products, $\Delta S^{\ominus}$ shows only a slight dependence on temperature. On this basis, Equation (8.16) means that a plot of ln K against reciprocal temperature should give a (tolerably) straight line, from whose slope we may derive $\Delta H^{\ominus}$.

This relation, due to Van't Hoff, means that standard enthalpy changes can be measured without the involvement of any calorimetry, simply by measuring an equilibrium constant as a function of temperature. It also yields the standard entropy change.

If the equilibrium constant increases with increasing temperature, ln K will decrease with increase in the reciprocal temperature, so the Van't Hoff plot will have a negative slope. However, given the minus sign on the right-hand side of Equation (8.16), this means $\Delta H^{\ominus}$ is positive. Conversely, if the equilibrium constant decreases with rising temperature, $\Delta H^{\ominus}$ is negative.

— EXAMPLE 8.1

For the reaction,

$$2SO_3(g) \; \rightleftharpoons \; 2SO_2(g) + O_2(g)$$

the equilibrium constant K_p was found to vary with the temperature as follows:

T/K	800	900	1000	1100
K_p	0.00103	0.0234	0.290	2.50

Hence evaluate $\Delta G^{\ominus}$ at 1000 K, $\Delta H^{\ominus}$ and $\Delta S^{\ominus}$ for this reaction.

— SOLUTION

To evaluate the standard Gibbs energy change at 1000 K, all we need is the equilibrium constant at that temperature

$$\begin{aligned}\Delta G^{\ominus} &= -RT \ln K_p \\ &= -8.314 \text{ J K}^{-1} \text{ mol}^{-1} \times 1000 \text{ K} \ln (0.290) \\ &= +10.3 \times 10^3 \text{ J mol}^{-1} \\ &= +10.3 \text{ kJ mol}^{-1} \end{aligned}$$

(8.17)

For the standard enthalpy change, we need to plot ln K_p against T^{-1}, as we do in Figure 8.4.

$T^{-1}/10^{-3}\text{K}^{-1}$	1.250	1.111	1.000	0.909
ln K_p	−6.88	−3.76	−1.24	0.92

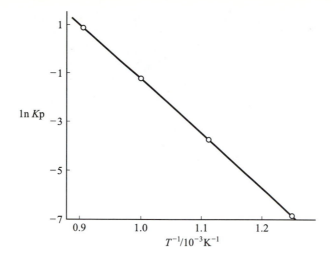

Figure 8.4 Plot of ln K_p against T^{-1} for the data given in Example 8.1.

This yields a good linear plot, whose slope is $-\Delta H^{\ominus}/R$.

$$\text{Slope} = \frac{-5.75 - 1.12}{(1.2 - 0.9) \times 10^{-3}\ \text{K}^{-1}} = -22.9 \times 10^3\ \text{K}$$

$$\Delta H^{\ominus} = -R \times \text{slope}$$
$$= -8.314\ \text{J K}^{-1}\ \text{mol}^{-1} \times -22.9 \times 10^3\ \text{K} \tag{8.18}$$
$$= +190.4 \times 10^3\ \text{J mol}^{-1}$$
$$= +190.4\ \text{kJ mol}^{-1}$$

To evaluate $\Delta S^{\ominus}$, let us substitute into Equation (8.16) at 1000 K.

$$\Delta S^{\ominus} = R \ln K_p + \frac{\Delta H^{\ominus}}{T}$$
$$= 8.314\ \text{J K}^{-1}\ \text{mol}^{-1} \times (-1.238) + \frac{190.4 \times 10^3\ \text{J mol}^{-1}}{1000\ \text{K}} \tag{8.19}$$
$$= 180.1\ \text{J K}^{-1}$$

From the tabulated data on the standard enthalpies of formation of SO_2 and SO_3, and for the standard entropies of all three substances, all at 298 K, the values obtained for $\Delta H^{\ominus}$ and $\Delta S^{\ominus}$ are only slightly different from those calculated above.

8.6 Effects of pressure on chemical equilibrium in the gas phase

The question of what effect the total pressure of a gaseous system has on the equilibrium is easily answered. Very simply, the equilibrium constant, K_p, is unaffected by changes in the partial pressures. Notwithstanding that, for some equilibria, the extent of the reaction that can take place at a fixed temperature is a function of the pressure.

To illustrate this, let us look at the behaviour of NO_2, which can combine with itself to form the **dimer**, N_2O_4.

$$2NO_2 \quad \rightleftharpoons \quad N_2O_4 \tag{8.20}$$

At 298 K, $\Delta G^{\ominus}$ for Equation (8.20) is -5.4 kJ mol^{-1}, which means that K_p is 8.84. So, if we have this system at 298 K and a total pressure of 1 bar, we may evaluate the extent to which it is dimerised.

It is useful to introduce the parameter α which we will call the degree of association. If we assume, hypothetically, that we add only NO_2 to this system, we use α to denote the fraction of this NO_2 which has dimerised. If the partial pressure of the 'original' NO_2 would have been P, then at equilibrium the partial pressures of NO_2 and N_2O_4 are $P(1 - \alpha)$ and $P\alpha/2$, which must add up to 1 bar.

For reaction (8.20) we have:

$$K_p = \frac{P_{N_2O_4}}{(P_{NO_2})^2} = 8.84 \tag{8.21}$$

Substituting into this equation, and using the fact that $P(1 - \alpha/2)$, the sum of the partial pressures of NO_2 and N_2O_4, is equal to 1 bar, we have a quadratic equation. One of the roots exceeds 1 and has no physical significance. The other is $\alpha = 0.835$, which gives $P_{NO_2} = 0.284$ and $P_{N_2O_4} = 0.715$ bar. These values add up to 1 and also they satisfy Equation (8.21).

If, on the other hand, we have a total pressure of 10^{-3} bar, then the value of α which is found to satisfy the equation is totally different at 0.0172. This gives $P_{NO_2} = 9.91 \times 10^{-4}$ and $P_{N_2O_4} = 8.7 \times 10^{-6}$ bar, values which are consistent with the same value for K_p in Equation (8.21), but which correspond to a vastly reduced extent of dimerisation of NO_2.

The significant point about the above example is that we have been considering a reaction in which there are different numbers of molecules on the two sides of the stoichiometric equation. At lower partial pressures, the equilibrium adjusts in such a way that a greater proportion of the matter present is to the side of the equation with the greater number of molecules. Similarly, an increase in the partial pressures would favour the side with the lesser number of molecules. For a reaction like,

$$H_2 + Cl_2 \quad \rightleftharpoons \quad 2HCl \tag{8.22}$$

with the same number of molecules on each side of the balanced equation, a change in the partial pressures has no such effect.

The discussion above of the NO_2/N_2O_4 equilibrium at reduced pressures was put in terms of these gases only being present. Actually, it is only the partial pressures of NO_2 and N_2O_4 which matter, so that we would have virtually the same extent of dimerisation if sufficient NO_2 to give a pressure of 10^{-3} bar were added to an atmosphere of air at 298 K.

One example of a practical application of the foregoing arises in the commercial synthesis of ammonia from its elements over a transition metal catalyst.

$$N_2 + 3H_2 \rightleftharpoons 2NH_3 \tag{8.23}$$

Since there are less molecules to the right-hand than to the left-hand side of this equation, increasing the pressure of the reacting system causes an increase in the proportion of the N atoms within it which may be present as NH_3 rather than as N_2.

8.7 The formation of metal complexes

In the absence of any stronger ligand, a metal ion in aqueous solution usually has water molecules complexed to it. When a better ligand is introduced, it will tend to displace these. A number of equilibria are then set up, between the complex with $(n-1)$ ligands and that with n, and for each of these there is an equilibrium constant, which we will label K_n. For example, for the complexing of Cu^{2+} ions by NH_3, we have:

$$
\begin{aligned}
[Cu(OH_2)_4]^{2+} + NH_3 &\rightleftharpoons [Cu(OH_2)_3NH_3]^{2+} + H_2O && K_1 \\
[Cu(OH_2)_3NH_3]^{2+} + NH_3 &\rightleftharpoons [Cu(OH_2)_2(NH_3)_2]^{2+} + H_2O && K_2 \\
[Cu(OH_2)_2(NH_3)_2]^{2+} + NH_3 &\rightleftharpoons [Cu(OH_2)(NH_3)_3]^{2+} + H_2O && K_3 \\
[Cu(OH_2)(NH_3)_3]^{2+} + NH_3 &\rightleftharpoons [Cu(NH_3)_4]^{2+} + H_2O && K_4
\end{aligned}
$$

Normally, the values of these equilibrium constants decline with increasing n, that is, $K_1 > K_2 > K_3 > K_4$. This trend does not imply that the third or the fourth NH_3 ligand is more weakly bonded to the metal ion than the first: rather, it reflects the fact that, whereas the first NH_3 ligand had the choice of four sites, the fourth has only one option available.

Some ligands are so constituted that they possess more than one atom with a lone pair capable of being donated to bonding orbitals involving the metal ion. If the geometry is suitable or the ligand molecule is sufficiently flexible, there is then the possibility that the ligand may be attached to the metal at two or more points. Such complexes are known as chelates and are quite abundant in nature.

The complexing of multidentate ligands (those capable of attaching at several points) to the metal ion appears, at first sight, to be more extensive than one might expect on the basis of the behaviour of comparable unidentate ligands. Qualitatively, we may consider the contrast between formate ion, HCO_2^-, and oxalate dianion, $^-O_2CCO_2^-$, as ligands. The first has negligible capacity for this role, whereas the second forms strong complexes, with the formation of 5-membered rings:

For example the ion Fe^{3+} complexes with three oxalate ions to form the ion $[Fe(C_2O_4)_3]^{3-}$.

A more quantitative comparison may be made by looking at the replacement of two water molecules around the Cu^{2+} ion either by two ammonia molecules or by one molecule of the bidentate ligand, ethylenediamine (1,2-diaminoethane). For the first, the equilibrium constant is 4×10^7 whereas for the second it is 4×10^{10}.

The accepted explanation of this apparent anomaly is given the name of the chelate effect. The attachment of the first end of a bidentate ligand to the metal ion is assumed to be attended by similar values (both negative) of ΔH and ΔS as for a unidentate ligand, leading to a similar value for the equilibrium constant. For the attachment of the second end of the bidentate ligand, while ΔH would be similar, ΔS should be appreciably greater, since the ligand has already surrendered its freedom of movement. Thus, ΔG for the attachment of the second end of a bidentate ligand will be much more negative than for a second unidentate ligand and this difference should be reflected in the relative values of the respective equilibrium constants.

The force of this argument is not limited to ligands that bind at two points, and multidentate ligands are quite common in nature. A simple example of a hexadentate ligand is ethylenediaminetetraacetate (EDTA),

which can attach at each N atom and at each carboxylate. For example, it binds strongly to the divalent ions of calcium and magnesium.

Suggested reading

LOGAN, S. R., 1988, Entropy of mixing and homogeneous equilibria, *Education in Chemistry*, **25**, 44–6.

GERHARTL, F. J., 1994, The A + B = C of chemical thermodynamics, *Journal of Chemical Education*, **71**, 539–48.

DAVID, C. W., 1988, An elementary discussion of chemical equilibrium, *Journal of Chemical Education*, **65**, 407–9.

Problems

8.1 Water vapour has been passed into the gases obtained by burning coke in a restricted supply of oxygen, so that equilibrium is achieved, at 500 K, in the water–gas shift reaction:

$$CO + H_2O \quad \rightleftharpoons \quad CO_2 + H_2$$

The partial pressures of the emerging gases were measured as follows: H_2, 332; CO_2, 337; H_2O, 81.4; CO, 10.6 torr. Evaluate K_p and $\Delta G^{\ominus}$ at for this reaction at 500 K. (1 bar 750 torr.)

8.2 For the gas phase reaction,

$$CO + Cl_2 \quad \rightleftharpoons \quad COCl_2$$

the value of $\Delta G^{\ominus}$ at 298 K is -73.2 kJ mol^{-1}. Evaluate K_p for this reaction at this temperature.

In this process, two poisonous gases are converted into an even more dangerous one. What pressures (assumed equal) of CO and Cl_2 are necessary in order that the equilibrium concentration of $COCl_2$ at 298 K will exceed that of each of the other gases?

8.3 For the synthesis of ammonia,

$$N_2 + 2H_2 \quad \rightleftharpoons \quad 2NH_3$$

the equilibrium constant K_p varies with temperature as follows:

T/K	623	673	723	773
K_p/bar^2	7.1×10^{-4}	1.66×10^{-4}	4.3×10^{-5}	1.45×10^{-5}

Hence evaluate the standard enthalpy change, $\Delta H^{\ominus}$, of this reaction and the standard entropy change, $\Delta S^{\ominus}$.

Whereas the commercial synthesis of ammonia is conducted at about 450°C, nitrogen fixation by legumes is accomplished at ambient temperatures. Evaluate the equilibrium constant for the above process at 300 K. Is this value more favourable or less favourable for ammonia synthesis than that at 723 K?

8.4 The protein myoglobin (M), present in muscle, can absorb or complex with oxygen:

$$M + O_2 \quad \rightleftharpoons \quad M.O_2$$

The extent to which myoglobin is present in the oxygenated form, $M.O_2$, rises with the partial pressure of O_2 in equilibrium with the dissolved oxygen, and tends towards 1 at high pressures. This ratio achieves 0.5 when the partial pressure of O_2 is 5.0 torr. Given that the Henry's Law constant of O_2 in water is 3.9×10^7 torr at 37°C, evaluate K_c for the above equilibrium at this temperature.

8.5 In ethanol, the equilibrium constant for the association process,

$$I_2 + I^- \quad \rightleftharpoons \quad I_3^-$$

is 1.4×10^4 dm^3 mol^{-1}. If we have an ethanolic solution of I_2, what molar concentration of KI should be added so that 80% of the 'iodine' is present as I_3^- ion (i.e. $[I_3^-] = 4[I_2]$)?

9

Solutions of electrolytes

One of the most important ideas to be introduced into chemistry was the notion that, in polar solvents, some compounds would ionise, partially or almost completely. Even water itself partially dissociates into the ions, H^+ and OH^-. Acids dissociate to produce the first and bases the second of these. A solution containing ions will conduct electricity, so the electrical conductivity is the appropriate parameter for monitoring the behaviour and the concentrations of ions in solution.

Since the dissociation of water is an equilibrium process, the product of the concentration of H^+ and OH^- is constant. This basic fact permits the preponderance of H^+ or of OH^- in a solution to be expressed quantitatively by the parameter pH. We look at systems in which the addition of an acid or a base effects a sharp change in the pH and also those, called buffer solutions, in which the pH change is minimal. For substances possessing more than one acidic or basic group, the nature of the species present changes with the pH. In the case of amino acids, this is a most important aspect of their behaviour.

9.1 Electrolytic dissociation

During the second half of the nineteenth century the store of chemical knowledge grew at a considerable rate. One feature then under study was the fact that solutions of certain substances, in solvents such as water, are capable of conducting electricity to an extent far in excess of the ability of the solvent itself. In the 1880s, a young man started his researches on this topic at Stockholm in Sweden, and reached a startling conclusion. In his doctoral thesis, Arrhenius proposed that when these substances were dissolved, heterolytic dissociation occurred, generating charged species called ions. The thesis was rated very lowly by the examiners, but now their names are obscure while that of the candidate is a scientific household word.

Salts, such as NaCl, which we met in Chapter 3, are examples of electrolytes. In solution, the ions Na^+ (a cation) and Cl^- (an anion) move around independently of each other, so that for each 'molecule' there are two entities present in solution. The need for the Van't Hoff i factor, mentioned in Chapter 7, is explicable on that basis.

The other important classes of electrolytes are acids and bases. An alkali metal hydroxide, like NaOH, is basically (no pun intended) analogous to a salt and leads to Na^+ and OH^- ions in solution. With many acids, there exists an individual molecule, which is virtually undissociated in the liquid state of the pure acid, but which undergoes heterolytic dissociation in a polar solvent:

$$HNO_3 \rightleftharpoons H^+ + NO_3^- \qquad (9.1)$$

Some acids, like nitric acid, are almost completely dissociated in dilute aqueous solution.

The question as to why these substances, salts, acids and bases, should behave in this way is, as before, answered in terms of the consequent lowering of the Gibbs energy of the system. In this connection, three points are important. Firstly, where a dissociation process occurs, producing more entities than were formerly present, the entropy will increase. Secondly, the process of heterolytic dissociation is assisted by having a solvent which substantially diminishes the Coulombic forces between ions, so that the attractive or cohesive forces between an anion and a cation are reduced far below the value in a vacuum at the same separation. In this respect, water is one of the very best solvents and in its presence the enthalpy change for the dissociation process is decreased. However, its role is not a totally passive one, in that the water molecules, having an excess of electronic charge on the O atom and a shortfall on the H atoms, behave as dipoles which tend to orient themselves around each ion, as sketched in Figure 9.1. This process is known as solvation.

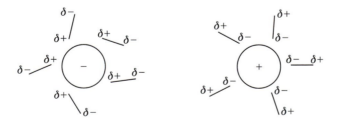

Figure 9.1 A schematic illustration of the solvation of anions and cations by the molecules of a polar solvent.

9.2 The dissociation of water and the pH scale

A significant property of water is its own heterolytic dissociation:

$$H_2O \quad \rightleftharpoons \quad H^+ + OH^- \qquad (9.2)$$

Since the molecule undergoing dissociation here is the solvent, we use as the dissociation constant the product of the concentrations of the species on the right-hand side, sometimes called the ionic product K_w:

$$K_w = [H^+][OH^-] \qquad (9.3)$$

At 298 K, K_w has the value of 1×10^{-14} mol^2 dm^{-6}, so that for pure water, $[H^+] = [OH^-] = 10^{-7}$ mol dm^{-3}.

A comment is perhaps desirable on the way that Equation (9.2), and even (9.1), have been written. If one takes an H atom and removes an electron to obtain H^+, all that is left is the nucleus. Does this minute ion really exist? The answer is that H^+, like any other ion, undergoes solvation, and because it is so small it interacts very strongly with the molecules of the aqueous solvent.

Some authors respond to the hydrated nature of H^+ by writing it as H_3O^+. This may be viewed as $H^+(H_2O)$. Detailed studies of aqueous acids do not tend to support the view that H^+ interacts in a unique way with one solvent molecule. Rather, they would argue for the structure, $H^+(H_2O)_n$, where n is variable but is around 4. Clearly, H^+ is hydrated just as are all other cations, and it may as well be written as H^+. If one wants particularly to emphasise the hydration, one can write it as H_{aq}^+.

If one adds to pure water some HNO_3, which dissociates in accordance with Equation (9.1) to produce H^+, the reverse reaction of Equation (9.2) will take place so that the equality in (9.3) is observed. As a result, the concentration of OH^- will now be less than 10^{-7} mol dm^{-3}. If KOH were added, there would be a similar consequence giving a concentration of H^+ much less than 10^{-7} mol dm^{-3}. Reaction (9.2) is a process which occurs very readily and quickly and it may be assumed that Equation (9.3) is always satisfied.

This means that the concentration of H^+ is a reliable indicator of the acidity or alkalinity of a solution. Whereas in neutral water it has the value of 10^{-7} mol dm^{-3}, in 0.1 mol dm^{-3} aqueous HNO_3 solution it is 10^{-1} mol dm^{-3}, and in 0.1 mol dm^{-3} aqueous NaOH solution it is 10^{-13} mol dm^{-3}. It would clearly be appropriate to use a logarithmic scale and the one employed is called the pH scale, defined as follows:

$$\begin{aligned} pH &= -\log_{10}[H^+] \\ &= \log_{10}\frac{1}{[H^+]} \end{aligned} \qquad (9.4)$$

Thus the pH values of the above three solutions are respectively 7 (the mid-point of the scale), 1 and 13. This 'definition' is not exact, but we have delayed introducing the concept of an activity coefficient until later.

9.3 The dissociation of weak acids

Carboxylic acids, such as ethanoic (acetic) acid, exemplify a class of substances which are only partially dissociated in aqueous solution:

$$RCO_2H \quad \rightleftharpoons \quad RCO_2^- + H^+ \tag{9.5}$$

Such are known as weak acids, in contrast to those such as HCl or HNO_3 which are completely or almost completely dissociated in a moderately con-centrated aqueous solution and are referred to as strong acids. For the dis-sociation process of this nature, the equilibrium constant is usually called K_a, since it pertains to an acid:

$$K_a = \frac{[RCO_2^-][H^+]}{[RCO_2H]} \tag{9.6}$$

The value of K_a depends on the identity of R but, typically, it may lie in the range 10^{-4} to 10^{-7}.

It is useful to introduce the parameter α to denote the degree of dissociation of an acid. If one has a solution of a weak acid, HA, of nominal concentration c, then the actual concentrations of A^- and of H^+ are both $c\alpha$ and that of HA is $c(1 - \alpha)$. (The term 'nominal concentration' is used to denote the hypothe-tical concentration that would be obtained if no dissociation occurred. That is, it is the quotient when the number of moles put into the solution is divided by the volume of that solution.) Thus we have the equality:

$$
\begin{aligned}
K_a &= \frac{[H^+][A^-]}{[HA]} \\
&= \frac{(c\alpha)^2}{c(1 - \alpha)} \\
&= \frac{c\alpha^2}{(1 - \alpha)}
\end{aligned}
\tag{9.7}
$$

Whereas K_a is a constant, independent of the concentration, α is not. If we assume a typical value for K_a, we may show that α tends to 1 as c becomes smaller, but as c is increased, α may become extremely small, as is illustrated in Figure 9.2.

However, although α falls as the nominal concentration c is increased, the concentration of H^+, equal to $c\alpha$, increases monotonically with increasing c, so that the pH is the lower the higher the nominal concentration. A measurement of the H^+ concentration thus permits the dissociation constant to be evaluated.

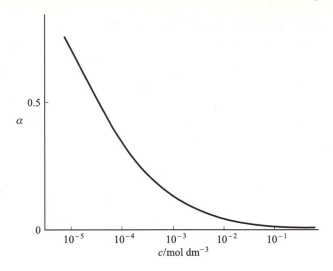

Figure 9.2 The variation of the degree of dissociation, α, of ethanoic acid ($K_a = 1.78 \times 10^{-5}$) with the nominal concentration, c.

— **EXAMPLE 9.1**

An aqueous solution of methanoic (formic) acid of nominal concentration 0.05 mol dm^{-3} was found to have a pH of 2.54. Evaluate the dissociation constant of methanoic acid.

— **SOLUTION**

$$pH = 2.54$$
$$\therefore [H^+] = 10^{-2.54} \tag{9.8}$$
$$= 2.88 \times 10^{-3} \text{ mol dm}^{-3}$$

$$\alpha = \frac{[H^+]}{c}$$
$$= \frac{2.88 \times 10^{-3}}{0.05} \tag{9.9}$$
$$= 0.0577$$

$$\therefore K_a = \frac{(c\alpha)^2}{c(1-\alpha)} = \frac{(2.88 \times 10^{-3})^2}{0.05 \times (1 - 0.0577)} \tag{9.10}$$
$$= 1.76 \times 10^{-4}$$

9.4 The neutralisation of a weak acid by a strong base

Suppose we have 50 cm^3 of an aqueous solution of 0.1 mol dm^{-3} HCl, to which we add small amounts of 0.1 mol dm^{-3} KOH solution. In the light of equilibrium (9.2), the addition of KOH must lead to the reaction of OH$^-$ with H$^+$ to give water, and the consequent depletion of the concentration of H$^+$. Initially we had (0.1 × 0.05) moles of H$^+$, but as x cm^3 of KOH is added, some of this is consumed leaving only (0.005 − 0.001x) moles in a volume of (50 + x)/1000 dm^3. The consequent increase in the pH with increasing x calculated in this way is shown in Figure 9.3.

The rationale of the previous paragraph applies only when the solution is acidic. When the number of moles of KOH added exceeds that of the number of moles of HCl present initially, we then have an alkaline solution containing rather more KCl. We now have (0.001x − 0.005) moles of KOH in a volume of (50 + x)/1000 dm^3, and on this basis the pH of the solution can be evaluated, as shown in Figure 9.3.

If the initial solution is that of a weak acid, the situation is slightly different. There are now two equilibria that need to be satisfied, namely those of Reactions (9.2) and (9.11):

$$HX \quad \rightleftharpoons \quad H^+ + X^- \tag{9.11}$$

The latter has now become the more important equilibrium and we may use it to obtain an expression for the pH:

$$K_a = \frac{[H^+][X^-]}{[HX]} \tag{9.12}$$

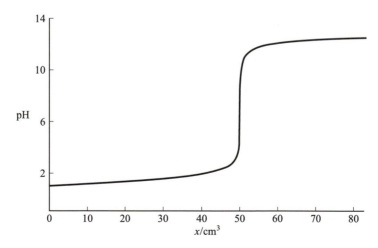

Figure 9.3 The variation of the pH of 50 cm^3 of a solution of 0.1 mol dm^{-3} aqueous HCl as a function of the volume, x, of 0.1 mol dm^{-3} aqueous KOH solution added.

$$\therefore [\text{H}^+] = K_a \frac{[\text{HX}]}{[\text{X}^-]} \tag{9.13}$$

Taking logarithms of each side and multiplying across by minus one, we have,

$$\text{pH} = pK_a + \log_{10}\left\{\frac{[\text{X}^-]}{[\text{HX}]}\right\} \tag{9.14}$$

where we have used pK_a to denote $-\log_{10}K_a$. This relation is known as the Henderson–Hasselbalch equation.

This equation is true even where there has been no neutralisation of the weak acid, but it becomes more useful during the neutralisation process. As KOH solution is added, OH^- reacts with H^+, so that additional X^- ions are formed, of an amount equal to that of the KOH added. Bearing in mind the very slight extent of dissociation of HX at a moderate concentration, as shown in Figure 9.2, in the partially neutralised solution one may, to a good approximation, equate the amount of X^- to the amount of KOH added. The ratio of the two concentrations in the last term of Equation (9.14) is the ratio of the number of moles of the species concerned. Since the nominal amount of X^- present is equal to that of the KOH solution added, the ratio $[\text{X}^-]/[\text{HX}]$ is equal to $x/(50-x)$. On this basis, the curve of pH against x can be calculated and this is shown in Figure 9.4.

When the amount of KOH added exceeds that of HX present initially, the latter has all been neutralised. Strictly speaking, equations (9.12) and (9.14) are still applicable, but the approximation referred to in the previous paragraph is no longer acceptable. In this regime, the pH has been estimated in the same

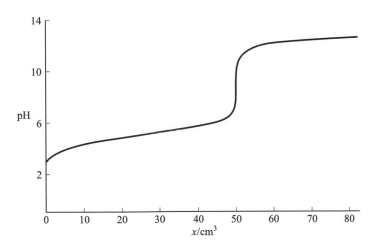

Figure 9.4 The variation of the pH of 50 cm³ of a solution of 0.1 mol dm⁻³ aqueous HX ($K_a = 10^{-5}$) as a function of the volume, x, of 0.1 mol dm⁻³ KOH added.

way as it was for Figure 9.3, treating the solution as one of KOH in the presence of a salt.

9.5 Buffer solutions

Referring to the diagrams in the previous section, it may be seen that whereas in Figure 9.3 the pH rises at an increasing rate over the range, $0 < x < 50$, in Figure 9.4 the pH curve shows a point of inflexion at $x = 25$, at which point the rate of increase is a minimum. Also, in the weak acid–strong base case, the rate of increase of pH with increasing volume of added KOH is very much less.

  This means that a partially neutralised solution of a weak acid is a system which undergoes very little change of pH when a small amount of a strong acid is added. The same would be true of adding a strong base, which would cause the pH to move along the curve in the reverse direction. A system like this which resists changes in pH is called a buffer solution, on the basis of the analogy of the mechanical device which brings a train to a halt at the end of the tracks. Clearly, a buffer solution is at its most effective when half of the weak acid has been neutralised. At that point, $[X^-] = [HX]$, and in consequence the pH of the solution is equal to the pK_a of the weak acid. The system can aim to be an effective buffer over a range of about one pH unit on either side of this figure.

—— **EXAMPLE 9.2** ————————————————————————

Given 0.1 mol dm^{-3} aqueous solutions of ethanoic (acetic) acid and of NaOH, along with normal volumetric apparatus, how would you produce buffer solutions of pH 4.40 and 5.35? The dissociation constant, K_a, of ethanoic acid is 1.78×10^{-5}.

—— **SOLUTION** ————————————————————————————

Let us assume we start off with 50 cm^3 of the ethanoic acid. The Henderson-Hasselbalch equation then becomes:

$$pH = pK_a + \log_{10}\left(\frac{x}{50 - x}\right) \tag{9.14a}$$

where x cm^3 is the volume of 0.1 mol dm^{-3} NaOH added. We obtain the value of pK_a by taking $\log_{10}$ of $(1.78 \times 10^{-5})^{-1}$. We then have:

(i) $4.40 = 4.75 + \log_{10}\left(\dfrac{x}{50 - x}\right)$

  $\therefore \dfrac{x}{50 - x} = 10^{-0.35} = 0.447$

  This gives a linear equation for x, whose solution is

  $x = 15.4$

So, to produce the buffer of pH 4.40, add 15.4 cm³ of NaOH to 50 cm³ of ethanoic acid.

(ii) $5.35 = 4.75 + \log_{10}\left(\dfrac{x}{50 - x}\right)$

$\therefore \dfrac{x}{50 - x} = 10^{0.60} = 10^{0.60} = 3.981$

$\therefore x = 40.0$

To produce the buffer of pH 5.35, add 40.0 cm³ of NaOH to 50 cm³ of ethanoic acid.

The form of Equation (9.14) implies that it is only the *ratio* of the concentrations of A^- and HA that is important, and that the actual values of these concentrations are irrelevant. This is true as an approximation, but it is acceptable only if the concentrations of the salt and the acid are sufficiently high that Equilibrium (9.11), the dissociation of HX, predominates over Equilibrium (9.2), the dissociation of the solvent. Clearly, if the concentrations of X^- and HX were both 10^{-9} mol dm^{-3}, one would expect a pH value close to 7, regardless of the pK_a value, since in this instance, with an extremely low concentration of electrolyte, the dissociation of water would be dominant. This means that a buffer solution resists changes to the pH not merely on adding acid or alkali, but also when diluting with water. However, in all cases there are limits to the buffering capacity of the solution. Equation (9.14) also runs into difficulties where the dissociation constant of the acid is extremely low, because the extent of the dissociation of water is not then negligible in relation to that of the acid.

9.6 Other weak electrolytes

Some electrolyte systems behave in a rather different way from Equation (9.11). We will first look at acids with more than one ionisable hydrogen atom and then at the behaviour of weak bases.

Butanedioic (succinic) acid, $HO_2C(CH_2)_2CO_2H$, has two acidic functional groups (i.e., it is a dibasic acid) and can undergo two dissociation processes,

$$
\begin{aligned}
(CH_2)_2\!\!\begin{array}{l} \diagup CO_2H \\ \diagdown CO_2H \end{array} &\rightleftharpoons (CH_2)_2\!\!\begin{array}{l} \diagup CO_2^- \\ \diagdown CO_2H \end{array} + H^+ \\[2em]
(CH_2)_2\!\!\begin{array}{l} \diagup CO_2^- \\ \diagdown CO_2H \end{array} &\rightleftharpoons (CH_2)_2\!\!\begin{array}{l} \diagup CO_2^- \\ \diagdown CO_2^- \end{array} + H^+
\end{aligned} \tag{9.15}
$$

155

to which we will attribute the dissociation constants K_{a1} and K_{a2}. One would expect the second to have the smaller value, since the species from which the second H^+ is dissociating already has a negative charge, albeit one which is formally attributed to the other end of the molecule. Experimentally, it is found that $K_{a1} = 6.89 \times 10^{-5}$ and $K_{a2} = 2.47 \times 10^{-6}$, that is, the first of the dissociation constants is greater than and the second is less than the value for ethanoic acid, for which $K_a = 1.78 \times 10^{-5}$.

When this acid is in an aqueous medium, the two equilibria corresponding to the above dissociation processes need to be satisfied simultaneously. Each equilibrium may be expressed as the corresponding Henderson–Hasselbalch equation and if we use, for short, the terms *succ* for $HO_2C(CH_2)_2CO_2H$, *succ$^-$* for $HO_2C(CH_2)_2CO_2^-$ and *succ^{2-}* for $^-O_2C(CH_2)_2CO_2^-$, then these become:

$$pH = pK_{a1} + \log_{10}\left\{\frac{[succ^-]}{[succ]}\right\} \tag{9.16}$$

$$pH = pK_{a2} + \log_{10}\left\{\frac{[succ^{2-}]}{[succ^-]}\right\} \tag{9.17}$$

From these we may deduce that at a pH equal to pK_{a1}, namely 4.16, $[succ] = [succ^-]$ and at a pH equal to pK_{a2}, namely 5.61, $[succ^{2-}] = [succ^-]$. It follows that we may identify the dominant species of the succinic acid system as a function of pH as follows:

pH range	Most abundant species
pH < 4.16	$HO_2C(CH_2)_2CO_2H$
4.16 < pH < 5.61	$HO_2C(CH_2)_2CO_2^-$
pH > 5.61	$^-O_2C(CH_2)_2CO_2^-$

The variations in the relative concentrations of these three species as a function of the pH are shown in Figure 9.5.

In addition to weak acids, there are also weak bases, of which aqueous ammonia is perhaps the most familiar. (Aqueous ammonia is often called ammonium hydroxide and written as NH_4OH, which is stoichiometrically the same as the left-hand side of Equation (9.18). There is no clear evidence to support the formula NH_4OH, which may be regarded as a convenient myth.) The ammonia molecule ionises with the assistance of a water molecule:

$$NH_3.H_2O \; \rightleftharpoons \; NH_4^+ + OH^- \tag{9.18}$$

The equilibrium constant to characterise this is normally written K_b, since it pertains to the dissociation of a base:

$$K_b = \frac{[NH_4^+][OH^-]}{[NH_3.H_2O]} \tag{9.19}$$

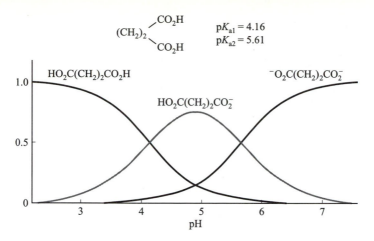

Figure 9.5 The relative abundances of the species, $HO_2C(CH_2)_2CO_2H$, $HO_2C(CH_2)_2CO_2^-$ and $^-O_2C(CH_2)_2CO_2^-$ as a function of the pH, given that the pK_a values of succinic acid are 4.16 and 5.61.

For ammonia, K_b has the value 1.77×10^{-5}. A solution of ammonia which has been partially neutralised by the addition of a strong acid will have similar properties to the partially neutralised solution of a weak acid discussed in the previous section. Rearranging Equation (9.19), we have:

$$[OH^-] = K_b \frac{[NH_3.H_2O]}{[NH_4^+]} \tag{9.20}$$

If we substitute for the OH^- concentration using Equation (9.3), we obtain, on rearranging:

$$\frac{1}{[H^+]} = \frac{K_b}{K_w} \cdot \frac{[NH_3.H_2O]}{[NH_4^+]} \tag{9.21}$$

Taking logarithms of both sides, we have:

$$pH = 14.0 - pK_b + \log_{10}\left\{\frac{[NH_3.H_2O]}{[NH_4^+]}\right\} \tag{9.22}$$

This demonstrates that if the weak base has been half neutralised, the pH of the solution is equal to 14.0 less the pK_b of the base. If the neutralisation has gone less than half-way, then the concentration of NH_3H_2O will exceed that of NH_4^+ so that the pH will be greater than this figure. A plot of pH against the extent of neutralisation is shown in Figure 9.6.

An alternative way to regard this matter is to treat NH_4^+ as an acidic species, which it can be claimed to be since it is capable of dissociating to give up H^+.

$$NH_4^+ + H_2O \rightleftharpoons NH_3.H_2O + H^+ \tag{9.23}$$

157

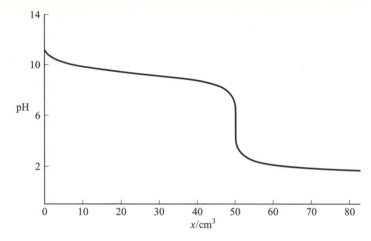

Figure 9.6 The variation of the pH of 50 cm^3 of 0.1 mol dm^{-3} aqueous ammonia solution as a function of the volume, x, of 0.1 mol dm^{-3} added.

For this, one may write the expression for the corresponding acid dissociation constant, which may be interrelated to K_b using equilibrium (9.2):

$$
\begin{aligned}
K_a &= \frac{[H^+][NH_3.H_2O]}{[NH_4^+]} \\
&= \frac{[H^+][NH_3.H_2O]}{[NH_4^+]} \cdot \frac{[OH^-]}{[OH^-]} \\
&= [H^+][OH^-] \cdot \frac{[NH_3.H_2O]}{[NH_4^+][OH^-]} \\
&= \frac{K_w}{K_b}
\end{aligned}
\tag{9.24}
$$

When the acid, NH_4^+, has been half-neutralised, one expects to have a pH equal to its pK_a value, which is clearly equal to $(14.0 - pK_b)$. This illustrates the mutuality in the behaviour of acids and bases.

9.7 Acid–base indicators

There exist a number of intensely coloured organic substances which undergo a protolytic dissociation process to produce an ion of a different colour. Almost invariably, these molecules have extensive delocalised π-electron systems. Their formulae need not concern us here, so we may refer to them as HE:

$$
HE \quad \rightleftharpoons \quad H^+ + E^-
\tag{9.25}
$$

Each of these will have an acid dissociation constant, which we will call K_{In}, and for each dissociation, the Henderson–Hasselbalch equation will be applicable:

$$pH = pK_{In} + \log_{10}\left\{\frac{[E^-]}{[HE]}\right\} \qquad (9.26)$$

Since both HE and E^- are present in the solution, the perceived colour depends crucially on their relative amounts. At a pH equal to pK_{In}, HE and E^- will have equal abundances. Arbitrarily, we may say that for the complete colour change, we will need to move from the situation where HE has 10 times the concentration of E^- to that where E^- has 10 times the concentration of HE. This requires a change of pH from ($pK_{In} - 1$) to ($pK_{In} + 1$), a range of two pH units. Salient details of a selection of acid–base indicators are listed in Table 9.1.

If one starts with a strong acid and adds to it a strong base, as was demonstrated in Figure 9.3, at the equivalence point (frequently called the end-point) the pH changes very sharply, over about 8 units of pH. To detect the end-point, one may use any indicator whose operational pH range lies within this band, for example methyl orange (3.2–4.4) or phenolphthalein (8.2–10.0). For either, a sharp colour change should take place on addition of the crucial drop of alkali.

When a strong base is added to a weak acid, the change in the pH at the end-point is rather less. For the parameters assumed in Figure 9.4, it is three or four units of pH. In choosing an indicator for this titration, one must now be more selective: phenolphthalein would suffice but methyl orange would not.

For the addition of a strong acid to a weak base, the fall in the pH at the end-point is of a similar magnitude, but it lies in a different range of pH, as shown in Figure 9.6. Once again, one must carefully select the indicator. In this instance, methyl red would be a good choice, but phenolphthalein a poor one.

If a weak base were added to a weak acid, the rise in pH at the end-point would be much more gradual than is shown in Figure 9.4. It would depend, of

Table 9.1 Some well-known acid–base indicators

Name	pH range	Colour change (acid to base)
Methyl orange	3.2–4.4	Orange to yellow
Metacresol purple	1.2–2.8	Red to yellow
	7.4–9.0	Yellow to purple
Thymol blue	1.2–2.8	Red to yellow
	8.0–9.6	Yellow to blue
Methyl red	4.4–6.2	Red to yellow
Chlorophenol red	5.2–6.8	Yellow to red
Phenolphthalein	8.3–10.0	Colourless to red
Alizarin yellow	10.1–12.0	Yellow to red

course, on the pK_a value of the acid and the pK_b value of the base, but in general it would not be possible to find a satisfactory indicator for this reaction. The reason is that the rise of two units of pH needed to accommodate the complete colour change would extend over the addition of too many cm^3 of the base solution. That is, the change in colour would be much too gradual and quite unsatisfactory.

In titrating a dibasic acid, one might choose an indicator that would change colour at the first neutralisation stage or one that would indicate the second. For the first stage, one would want an indicator whose operative pH range lay intermediate between the two pK_a values, and for the second, one changing colour a few pH units higher than the second pK_a value.

9.8 Acid-base equilibria in amino-acids

The simplest amino-acid is glycine, often written as $H_2NCH_2CO_2H$, which possesses both an acidic and a basic function. In solution, the carboxylic group dissociates to give $-CO_2^-$, and the amine group becomes protonated to yield $-NH_3^+$. The molecule, as written above, does not actually exist in solution, for the reason that the pH value at which the carboxylic group dissociates is well below that at which the amino group has acquired its proton. The equilibria are,

$$\overset{+}{H_3N}CH_2CO_2H \overset{-H^+}{\rightleftharpoons} \overset{+}{H_3N}CH_2CO_2^- \overset{-H^+}{\rightleftharpoons} H_2NCH_2CO_2^- \tag{9.27}$$

where the pK_a values are respectively 2.3 and 9.7. The species $\overset{+}{H_3N}CH_2CO_2^-$, carrying two formal charges but being overall electrically neutral, is called a zwitterion, and is the predominant species at pH values between 2.3 and 9.7.

A concept of some usefulness here is the isoelectric point, at which the concentration of the zwitterion is at its maximum and those of the positively and negatively charged species are equal. That is:

$$[\overset{+}{H_3N}CH_2CO_2H] = [H_2NCH_2CO_2^-] \tag{9.28}$$

These concentration terms appear in the respective expressions for the dissociation constants, K_{a1} and K_{a2}:

$$K_{a1} = \frac{[H^+][\overset{+}{H_3N}CH_2CO_2^-]}{[\overset{+}{H_3N}CH_2CO_2H]} \tag{9.29}$$

$$K_{a2} = \frac{[H^+][H_2NCH_2CO_2^-]}{[\overset{+}{H_3N}CH_2CO_2^-]} \tag{9.30}$$

Equating the concentration of $\overset{+}{H_3N}CH_2CO_2H$ obtained from Equation (9.29) with that of $H_2NCH_2CO_2^-$ from Equation (9.30), we have:

$$\frac{[H_3\overset{+}{N}CH_2CO_2^-][H^+]}{K_{a1}} = \frac{K_{a2}[H_3\overset{+}{N}CH_2CO_2^-]}{[H^+]} \tag{9.31}$$

The concentration of the zwitterion cancels out, leaving:

$$[H^+]^2 = K_{a1} \cdot K_{a2}$$
$$\therefore pH = \tfrac{1}{2}(pK_{a1} + pK_{a2}) \tag{9.32}$$

So, for glycine, the isoelectric point, pI, occurs at pH = (2.3 + 9.7)/2 = 6.0.

Of the other amino-acids of biological importance, some conform to the formula, $H_2NCHRCO_2H$, where R denotes an alkyl group, and their acid–base behaviour resembles that of glycine. Other amino-acids contain additional functional groups capable of engaging in acid–base reactions, and these have more than two pK_a values. As an illustration, we may look at the behaviour of the essential amino-acid, lysine, $H_2N(CH_2)_4CH(NH_2)CO_2H$.

$$
\begin{array}{ccccccc}
\overset{+}{N}H_3 & & \overset{+}{N}H_3 & & NH_2 & & NH_2 \\
| & & | & & | & & | \\
(CH_2)_4 & \overset{-H^+}{\rightleftharpoons} & (CH_2)_4 & \overset{-H^+}{\rightleftharpoons} & (CH_2)_4 & \overset{-H^+}{\rightleftharpoons} & (CH_2)_4 \\
| & & | & & | & & | \\
CH{-}CO_2H & & CH{-}CO_2^- & & CH{-}CO_2^- & & CH{-}CO_2^- \\
| & & | & & | & & | \\
{}^+NH_3 & & {}^+NH_3 & & {}^+NH_3 & & NH_2
\end{array}
\tag{9.33}
$$

The pK_a values are 2.2, 9.0 and 10.5. The four species involved carry, from the left, net charges of $2+$, $+$, zero and $-$. The isoelectric point is determined as the pH value at which the concentration of the third of these species is at a maximum and those of the second and the fourth are equal. Thus we have:

$$pI = \tfrac{1}{2}(pK_{a2} + pK_{a3}) \tag{9.34}$$
$$= 9.75$$

9.9 Non-ideal solutions and the activity coefficient

Ideal solutions have been introduced as solutions whose behaviour is analogous to that of a perfect gas. The relevant question is: how does the molar Gibbs energy vary with the concentration of the solute? If the manner in which G increases with the concentration, c, is the same as that of the increase of G with P for a perfect gas, then the solution is ideal.

As we have seen in Chapter 5, the equation of state for a perfect gas represents the behaviour to be expected if the gas molecules (i) have no finite size and (ii) exert no forces on each other except for mutual repulsion when two molecules are seeking to occupy the same position. Real gases meet neither criterion: they consist of molecules of finite size, which exert net forces of attraction when in close proximity.

The solutes whose behaviour one might most expect to deviate from ideality would belong to two classes:

(a) Those consisting of very large molecules, which would inevitably interfere with each other even in very low concentrations.

(b) Electrolytic solutions, because they contain ions whose forces of mutual interaction are both far stronger and of much longer range than the dipole–dipole or the dispersion forces between gas molecules.

Following a suggestion of G.N. Lewis, the molar Gibbs energy of the solute X in a non-ideal solution is described by the equation:

$$G_X = G_X^{\ominus} + RT \ln a_X \tag{9.35}$$

On the basis of the behaviour found experimentally, this equation serves as a definition of a, the activity of the solute. The parameter a is related to the concentration, c, by the equation $a = c\gamma$, where the activity coefficient, γ, is a function of the concentration. Clearly, for an ideal solution, γ is equal to 1 and usually the activity coefficient approaches this value as the solution is made more dilute. For non-ideal solutions, γ differs from unity: more often it is less than 1.

A theoretical treatment of electrolyte solutions was developed in the 1920s by Debye and Hückel. They recognised that around any negative ion there will be, statistically, a preponderance of positive ions, and vice versa. This preponderance, amounting to an average excess of ions of opposite sign to the extent of the charge on the central ion, is called the ionic atmosphere. The greater the electrolyte concentration, the closer it lies to the ion. Using the potential generated by this ionic atmosphere and making a reasonable assumption, it was possible to derive an expression for the activity coefficient which, for aqueous solutions at 298 K, is:

$$\log_{10} \gamma_X = -0.51 z_X^2 \sqrt{I} \tag{9.36}$$

where z_X denotes the number of electronic charges on ion X and I represents the ionic strength, given by:

$$I = \tfrac{1}{2} \Sigma(c_i z_i^2) \tag{9.37}$$

For a 1:1 electrolyte like NaCl, $I = c$, and at a concentration of 10^{-2} mol dm^{-3}, γ is evaluated as 0.89. As the charge on the ion is increased or the concentration is increased, the ionic interactions increase and γ becomes smaller. However, in view of the simplifications implicit in the Debye–Hückel theory, Equation (9.36) is increasingly unreliable as the ionic strength is further increased.

The important implication of Equation (9.35) is, of course, that our expressions for the equilibrium constants of ionic equilibria should include the activity coefficients of all the species involved. In the case of Reaction (9.2), if one is merely trying to evaluate the concentration of H$^+$ in pure water, then their omission does not much matter. However, in strong solutions of acids, bases, or salts, where the ionic strength is substantial, the activity coefficients will deviate considerably from unity. The correct equation is then:

$$K_w = [\text{H}^+]\gamma_{\text{H}^+}[\text{OH}^-]\gamma_{\text{OH}^-} \tag{9.3a}$$

9.10 The solubility of salts

Since salts dissociate on dissolution, the rules which govern their being present in solution are slightly different from those pertaining to non-electrolytes. At equilibrium, one may equate the Gibbs energy per mole of the solid and the solute. The latter consists of anions and cations, which implies a fixed value for the Gibbs energy of the anions plus that of the cations,

$$\text{MX}_n(\text{s}) \quad \rightleftharpoons \quad \text{M}^{n+} + n\text{X}^- \tag{9.38}$$

For any dissolved species, the Gibbs energy per mole depends on the concentration, so we have:

$$G^{\ominus}_{\text{MX}_n} = G^{\ominus}_{M^{n+}} + \text{RT}\ln[\text{M}^{n+}] + n\{G^{\ominus}_{\text{X}} + \text{RT}\ln[\text{X}^-]\} \tag{9.39}$$

Rearranging, we have:

$$G^{\ominus}_{\text{MX}_n} - G^{\ominus}_{M^{n+}} - nG^{\ominus}_{\text{X}^-} = -\text{RT}\ln\{[\text{M}^{n+}][\text{X}^-]^n\} \tag{9.40}$$

In this equation, the value of the left-hand side is fixed, and so must also be that of the right. This implies a fixed value for $[\text{M}^{n+}][\text{X}^-]^n$, which is called the solubility product. We will use the symbol, K_{sp}. This term tends to be applied only in regard to sparingly soluble salts, where the approximations implicit in Equations (9.39) and (9.40) are much more acceptable.

An example of a solubility product is the value of 2.4×10^{-5} mol^2 dm^{-6} for CaSO$_4$. This means that by adding this salt to water, one may attain a concentration equal to the square root of this figure, namely 4.9×10^{-3} mol dm^{-3}. However, if one were to add CaCl$_2$ solution to a 0.50 mol dm^{-3} aqueous solution of Na$_2$SO$_4$, then the solubility product of CaSO$_4$ would be attained once the concentration of Ca^{2+} reached 4.8×10^{-5} mol dm^{-3}, since the product of the concentrations of SO$_4^{2-}$ and Ca^{2+} has now risen to its limiting value. The addition of more CaCl$_2$ would lead to the precipitation of CaSO$_4$ until its solubility product was no longer being exceeded.

9.11 Electrical conductance by electrolytic solutions

If two electrodes of a non-corrosive metal such as platinum are dipped into an electrolyte solution and a voltage applied, then a current will flow between them. The magnitude of the current depends both on the area and on the separation of the electrodes, so a conductance cell of fixed geometry is normally used. Applying a DC voltage will lead to chemical reaction at the electrodes, so an AC voltage is used, with a frequency of perhaps 1000 Hz. Such a system obeys Ohm's Law, and the reciprocal of the electrical resistance, R, is a criterion of how well electricity is conducted between the electrodes. The specific conductivity, κ, is defined as:

$$\kappa = \frac{1}{R} \cdot K_{cell} \tag{9.41}$$

where the cell constant, K_{cell}, has the dimensions of reciprocal length, so that the units of κ are S cm^{-1}.

A pure solvent such as water gives an extremely low value for κ. However, when an electrolyte is added, the specific conductance rises by an amount which depends on the concentration. It is convenient to define the molar conductivity of a solution as,

$$\Lambda = \frac{\kappa - \kappa_0}{a} \tag{9.42}$$

where the numerator is the difference between the specific conductivity of the solution and of the solvent, and a denotes the concentration of the solute in mol cm^{-3}. Concentrations are frequently cited as mol dm^{-3}, and if this value is denoted by c, then we have,

$$\Lambda = \frac{1000(\kappa - \kappa_0)}{c} \tag{9.42a}$$

and the units of Λ are S cm^2 mol^{-1}.

When the concentration of an electrolyte is varied, there are two basic patterns for the consequent variation in Λ, as is demonstrated in Figure 9.7. As c increases, either Λ falls very slightly, as in case (a), or it falls quite

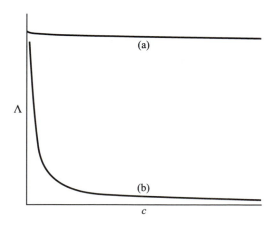

Figure 9.7 The patterns of variation of the molar conductivity, Λ, of an electrolyte as a function of the nominal concentration, c.

considerably, as in (b). The explanation of type (a) behaviour was provided by Arrhenius in 1884, in terms of the dissociation of an electrolyte, such as a salt, into ions which can move around independently and conduct electricity. The slight decrease in Λ with increasing concentration arises because the higher the concentration, the greater is the mutual interference between ions.

By extrapolating back to zero concentration, we obtain Λ_0, the molar conductivity of the electrolyte at infinite dilution. Sample values of Λ_0 (in S cm^2 mol^{-1}) are: LiBr, 117.1; KBr, 151.9; LiOH, 236.3; and KOH, 271.1. In each case, the molar conductivity of the hydroxide exceeds that of the bromide by 119.2. Also, for the hydroxides as for the bromides, the conductance of the potassium compound exceeds that of the lithium compound by 34.8. The pattern suggests that each ion has its own value, λ, for its conductivity at infinite dilution and that the molar value is the sum of the two individual components. This is, of course, the pattern to be expected if each ion moves around independently and responds to the applied field at its own characteristic rate.

In Table 9.2, the ionic molar conductances, in water at infinite dilution, of some anions and cations are listed. The means by which these are derived, for even one electrolyte, are beyond the scope of this work. It is notable that among cations, the ion with the greatest value is H^+, which conducts much better than any other cation. Likewise, the hydroxide ion easily tops the list of anions. The reason for this is related to the fact that these ions are obtained by the dissociation of the solvent. This means that they can migrate through the solvent by a mechanism involving only the movement of valence electrons, as illustrated below:

Table 9.2 Ionic molar conductances of some aqueous cations and anions at infinite dilution at 298 K

Cation	λ^+/S cm^2 mol^{-1}	Anion	λ^-/S cm^2 mol^{-1}
H^+	350	OH^-	198
Li^+	39	F^-	55.5
Na^+	51	Cl^-	75.5
K^+	74.5	Br^-	78
Cs^+	77	I^-	77
Ag^+	63.5	NO_3^-	70.5
NH_4^+	74.5	ClO_4^-	68
NMe_4^+	45	$CH_3CO_2^-$	41

In Figure 9.7, type (b) behaviour is found in the case of an electrolyte which is only partly dissociated. In effect, Λ varies with c in much the same way as does α, the degree of dissociation, so this curve closely resembles that in Figure 9.2.

9.12 Conductimetric titrations

If we have a solution containing several different ionic species, of varying concentrations, the conductivity κ (which at moderate concentrations is many times greater than that of the pure solvent) reflects the aggregate of their contributions to conducting electricity through the solution,

$$\kappa = \frac{1}{10^3}[c_1\lambda_1 + c_2\lambda_2 + \cdots c_n\lambda_n] \tag{9.43}$$

where the c_i values are the actual concentrations of each ionic species present. In acid–base titrations, as a titrant solution is added to the flask from a burette, the usual chemical consequence is that one ionic species is replaced in the solution by another. If the λ value of the new ion is different from that of the original, then κ should rise (or fall) as a linear function of the volume x of titrant added. Once the original ion has all been consumed, then further addition will produce a change of a different pattern, so that the plot of κ against x will show a pronounced break. In this way, the conductivity of the solution can be used to indicate stoichiometric equivalence.

Two qualifications are necessary. The first is that the concentration of the titrant should be appreciably greater than that of the solution being titrated, so that there is only a slight dilution effect accompanying additions from the burette. The second is that while values of λ for individual ions decrease slightly with increasing concentration, just as do Λ values for strong electrolytes, such variation is determined by the total ionic concentration (the ionic strength) rather than by that of the individual species. Since the former will remain essentially constant during the titration, this means that the various λ values for different ions will all remain related in much the same way to the respective limiting conductivity values.

Let us consider the addition to a fairly dilute (e.g. 0.01 mol dm^{-3}) solution of HCl, of a more concentrated (e.g. 0.1 mol dm^{-3}) solution of NaOH. If, initially, the concentration of the HCl is c, then we have:

$$\kappa = \frac{1}{10^3}[c(\lambda_{H^+} + \lambda_{Cl^-})] \tag{9.44}$$

Let us assume we add sufficient NaOH to produce a concentration c_1 in this volume, and that c_1 is less than c. The OH$^-$ will react quantitatively with the H$^+$, so that the conductivity is given by:

$$\kappa = \frac{1}{10^3}[(c - c_1)\lambda_{H^+} + c\lambda_{Cl^-} + c_1\lambda_{Na^+}]$$

$$= \frac{1}{10^3}[c(\lambda_{H^+} + \lambda_{Cl^-}) - c_1(\lambda_{H^+} - \lambda_{Na^+})]$$

$$(9.45)$$

Thus κ decreases linearly with x over the range, $0 < c_1 < c$.

Assume we have added sufficient NaOH to give a concentration c_2 in this volume, where $c_2 > c$. As a consequence of the reaction of the OH^- with H^+, the latter will be consumed, and the conductivity will now be given by:

$$\kappa = \frac{1}{10^3}[c\lambda_{Cl^-} + c_2\lambda_{Na^+} + (c_2 - c)\lambda_{OH^-}]$$

$$= \frac{1}{10^3}[c(\lambda_{Na^+} + \lambda_{Cl^-}) + (c_2 - c)(\lambda_{Na^+} + \lambda_{OH^-})]$$

$$(9.46)$$

So in the region where c_2 exceeds c, κ will rise linearly with x. This is demonstrated in Figure 9.8.

As was discussed in section 9.7, when a weak acid is titrated with a weak base, the change in the pH is so gradual that the end-point cannot be detected using an indicator. However, this may be achieved conductimetrically.

Let us start off with ethanoic acid (HAc), which is only slightly dissociated at moderate concentrations. Thus, the κ value is quite low. When aqueous ammonia is added, neutralisation of the acid takes place to generate ammonium ethanoate, a fully dissociated salt. So the specific conductivity rises steeply and linearly as x increases, since the common concentrations of NH_4^+ and Ac^- are proportional to x.

After the end-point has been attained, excess ammonia is being added to the salt. At moderate concentrations, this base is only very slightly dissociated. In

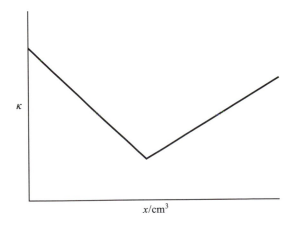

Figure 9.8 The variation of the specific conductivity, κ, of 50 cm^3 of 0.01 mol dm^{-3} aqueous HCl solution as a function of the added volume, x, of 0.1 mol dm^{-3} NaOH solution.

this instance, there are already NH_4^+ ions present so the extent of dissociation of the excess ammonia is minimal. Consequently, κ will decline slightly because of the dilution achieved by the extra solvent. There is thus a sharp break in the curve, to designate the end-point, as shown in Figure 9.9.

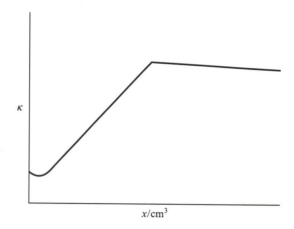

Figure 9.9   The variation of the specific conductivity, κ, of 50 cm³ of 0.01 mol dm^{-3} aqueous ethanoic acid as a function of the added volume, x, of 0.1 mol dm^{-3} aqueous ammonia solution.

9.13   Colloidal solutions

A colloid is a dispersion of small particles of one phase in another phase. The particles we are dealing with here are larger than normal molecules, with dimensions of up to a few hundred nanometres. So we may have small solid particles in a liquid (called a sol), particles of liquid or of solid in a gas (an aerosol) or of a liquid dispersed in another liquid (an emulsion). In the present context, it is the colloids involving a liquid dispersed phase that are of most interest.

The role of a liquid solvent in a colloidal system is not in general a passive one. There are usually strong interactions between the colloidal particles and the solvent, of one of two basic types. In some cases, such as finely dispersed metallic particles (metal sols), the colloid is lyophobic, which means solvent repelling. More often, there are attractive interactions between the colloid and the solvent, for which the term is lyophilic. An extreme example of this is a gel, in which a substance like gelatin interacts so strongly with the solvent molecules that all of the dispersed phase has been absorbed by the sol.

One significant point in relation to colloids is that a considerable proportion of the molecules of the dispersed phase are in contact with the solvent. For particles of 100 nm in diameter, this is only about 2%, but if the diameter is 10 nm, it has risen to about 23%. The behaviour of these molecules at the inter-face then becomes extremely important.

In thermodynamic terms, a colloid represents at best a metastable system, in which the Gibbs energy decreases the larger are the particles in the dispersed phase. For example, smoke particles aggregate and settle out on any available surface, and the small droplets of fat in milk tend to coalesce and rise to the surface as cream. There are various reasons why the inherent instability of colloids does not lead to much more rapid separation. In part it is due to the strong attractive intermolecular forces across the interface. Additionally, many colloid suspensions, particularly in polar liquids such as water, carry electric charges. Since the particles are all charged in the same sense, and so mutually repel, particle coagulation is thereby inhibited.

Where there is any electrolyte present in the solution, this interacts with a charged particle by forming an electrical double layer around it in the solvent. For example, if a particle is positively charged, then immediately around it there will be an excess of ions of negative charge, to the extent of the charge carried by the particle. This means that outside the electrical double layer, the charge on the particle is effectively shielded. The dimensions of the electrical double layer depend inversely on the ionic concentration: at 10^{-3} mol dm^{-3} its half-width is 6.7 nm, whereas at 10^{-1} mol dm^{-3} this has fallen to 0.7 nm. So the higher the salt concentration, the closer the particles need to be before they experience mutual repulsion on account of the like charges. In this way, the addition of a salt facilitates particle coalescence.

One illustration of the effect of the electrical double layer is provided by colloidal clay particles in a river. These may successfully be carried down-stream for hundreds of miles, but when the river reaches the sea, the increasing concentration of salts causes a massive contraction in the dimensions of the electrical double layer. This results in coagulation, leading to the deposition of silt in the estuary. Another example, on a less global scale, is enacted in the bathroom. A new razor blade pierces the skin and draws blood: a styptic pencil containing highly charged salts is applied and the presence of these ions in solution induces the rapid congealing of blood to staunch the flow.

Another approach to coagulation is to alter the solution in such a way that the particles become uncharged. For natural macromolecules like proteins, the charge depends on the pH in a more complex way than for a single amino-acid, but each has a pH at which the net charge is, on average, equal to zero. For myoglobin, this isoelectric point is at pH 7.0, whereas for egg albumin it is at pH 4.55. These values denote the condition where each protein is most readily precipitated.

Suggested reading

BRETT, C. M. A. and BRETT, A. M. O, 1993, *Electrochemistry: Principles, Methods and Applications*, Oxford: Oxford University Press.

ROBBINS, J., 1972, *Ions in Solution*, Oxford: Oxford University Press.

Problems

9.1 At a nominal concentration of 0.02 mol dm^{-3}, an acid HA has a degree of dissociation, α, equal to 0.053. Evaluate the acid dissociation constant, K_a, of HA.

9.2 The acid dissociation constant of lactic acid, $CH_3CH(OH)CO_2H$, is 1.5×10^{-4}. At a nominal concentration of 0.01 mol dm^{-3}, evaluate the degree of dissociation, α.

9.3 You are provided with 0.1 mol dm^{-3} solutions of propanoic acid ($K_a = 1.34 \times 10^{-5}$) and of NaOH. Describe how to make up a buffer solution of pH 5.25.

9.4 You are provided with 0.1 mol dm^{-3} solutions of propanoic acid ($K_a = 1.34 \times 10^{-5}$) and of sodium propanoate. Describe how you would make up a buffer solution of pH 4.75.

9.5 You are provided with 0.1 mol dm^{-3} solutions of oxalic acid and of NaOH. The dissociation constants of oxalic acid are 5.9×10^{-2} and 6.4×10^{-5}. Describe how to prepare a buffer solution of pH 4.50.

9.6 Given 0.1 mol dm^{-3} solutions of aqueous ammonia ($K_b = 1.77 \times 10^{-5}$) and of aqueous HCl, describe how to prepare a buffer solution of pH 9.35.

9.7 In aqueous solution, carbon dioxide becomes hydrated to give carbonic acid, H_2CO_3, of which the first pK_a value is 6.1. Also CO_2 is present in the bloodstream. What should be the ratio of the concentrations of carbonic acid, H_2CO_3, to bicarbonate anion, HCO_3^-, in blood at pH 7.4?

9.8 α-Tartaric acid, $HO_2CCH(OH)CH(OH)CO_2H$, has acid dissociation constants of 1.04×10^{-3} and 4.55×10^{-5}. What pH value should be sought in order to have the highest proportion of this compound in solution as the ion, $HO_2CCH(OH)CH(OH)CO_2^-$?

9.9 I have a solution containing both hydrochloric acid and ethanoic acid. I wish to titrate it with standard NaOH solution to assay the concentration of each. Is there an indicator listed in Table 9.1 which would serve this purpose?

9.10 The pK_a values of the amino acid tyrosine, $HOC_6H_4CH_2CH(NH_2)CO_2H$, are listed as 9.1 and 10.1 and the pK_b value as 11.8. Evaluate the isoelectric point of tyrosine.

9.11 The solubility product of manganese sulfide, MnS, is 1.4×10^{-5}. The dissociation constants of H_2S are 9×10^{-8} and 1.1×10^{-11}. In a solution, buffered to pH 4, containing 0.1 mol dm^{-3} MnSO$_4$, will 0.050 mol dm^{-3} H_2S cause the precipitation of MnS?

10

Electrochemical cells, redox reactions and bioenergetics

In 1791, a fortuitous discovery was made that if two different metallic conductors are dipped into solutions of electrolytes in close proximity, then an electric current may flow when an electrical connection is made between these two conductors. A significant aspect of this phenomenon relates to devices for the generation of electric currents, but electrochemical cells have also been devised to make measurements of appreciable chemical and clinical significance.

The electron transfer processes involved in electrochemical cells may serve to introduce the concepts of oxidation and reduction. These have considerable importance in a range of chemical and biochemical reactions, including those in which food is metabolised *in vivo*. In this connection, there is a most important substance known by the acronym ATP. Its crucial role in enabling certain difficult but necessary reactions to take place is explained in accordance with our simple criterion, that the Gibbs energy of a chemical system tends towards a minimum.

10.1 Galvanic cells

An example of a simple electrochemical cell is shown in Figure 10.1. This may adequately be written as:

$$Zn(s) \mid Zn^{2+}(aq) \parallel Cu^{2+}(aq) \mid Cu(s) \qquad (10.1)$$

This cell has two separate solutions, one of aqueous $CuSO_4$ and the other of aqueous $ZnSO_4$, joined by a salt bridge or a porous plug, which effects an electrical connection between them without permitting them to mix. The presence of liquid junctions introduces complications into the theoretical interpretation of cells, but in some instances their presence is unavoidable.

At an interface between each metal and a solution of its salt, the following process is a feasible one:

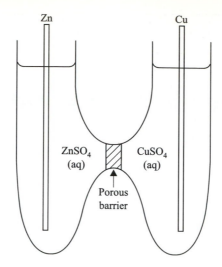

Figure 10.1   An illustration of a Daniell cell, with the solutions of $ZnSO_4$ and $CuSO_4$ separated by a porous barrier to prevent their mixing, but allowing an electrical connection between them.

$$M(s) \rightleftharpoons M^{2+}_{(aq)} + 2e^- \tag{10.2}$$

The metallic conductors would permit the removal of the electrons generated by the forward reaction or the furnishing of the electrons necessary for the reverse step. Moreover, since the two metallic conductors are electrically linked, for Reaction (10.2) to take place at one electrode it is necessary that the reverse process occurs at the other.

For the cell indicated in Equation (10.1), known as the Daniell cell, the chemical reaction is:

$$Zn(s) + Cu^{2+}_{(aq)} = Zn^{2+}_{(aq)} + Cu(s) \tag{10.3}$$

That is, metallic zinc goes into solution at the Zn electrode while at the other, copper is deposited. This occurs because it leads to a lowering of the Gibbs energy of the system.

The decrease in the Gibbs energy may also be equated to the maximum amount of electrical work that can be done as a consequence, which gives the relation,

$$-\Delta G = nFE \tag{10.4}$$

where n is the number of electrons, two in this instance, requiring to be transferred for the chemical reaction to take place, F denotes Faraday's constant and E is the electromotive force (e.m.f.) of the cell.

Two of the concepts already mentioned require more careful definitions. By the Stockholm convention, the electromotive force of a cell denotes the potential of the electrode written on the right with respect to that written on the left. Thus a cell e.m.f. has a sign, $+$ or $-$, which will be inverted if the cell is written in reverse order. The term 'cell reaction' is applied to the overall chemical process that takes place if positive electricity moves through the cell from left to right. Clearly, if the cell were to be written down in reverse order, the cell reaction would likewise be inverted so that the signs of each side of Equation (10.4) would be changed. For the cell specified in Equation (10.1), the cell reaction as defined by this convention is as given in Equation (10.3). The latter also represents the reaction which occurs spontaneously when an electrical connection is made between the electrodes, but this and the cell reaction coincide only if the right-hand electrode is the positive one.

10.2 The Nernst equation

For a chemical reaction, the standard Gibbs energy change, $\Delta G^{\ominus}$, represents the change in the Gibbs energy when the reactants, each in their standard states, are transformed to the products, in their respective standard states, for the number of moles of reactant indicated in the balanced equation for this reaction. More typically, neither the reactants nor the products are at unit activity so the actual Gibbs energy change is ΔG.

For a chemical reaction that is not at equilibrium, such as the hypothetical reaction,

$$A + B + . \quad \rightleftharpoons \quad P + Q + . \tag{10.5}$$

we have, similarly to Equation (8.10), but using activities rather than concentrations,

$$\Delta G = \Delta G^{\ominus} + RT \ln \frac{a_P \cdot a_Q \cdot}{a_A \cdot a_B \cdot} \tag{10.6}$$

where a_X denotes the activity of component X. As applied to the reaction occurring within an electrochemical cell, we may substitute using Equation (10.4):

$$-nFE = -nFE^{\circ} + RT \ln \frac{a_P \cdot a_Q \cdot}{a_A \cdot a_B \cdot} \tag{10.7}$$

Dividing across by $-nF$, this gives us the Nernst equation:

$$E = E^{\circ} - \frac{RT}{nF} \ln \frac{a_P \cdot a_Q \cdot}{a_A \cdot a_B \cdot} \tag{10.8}$$

In the Daniell cell, the activities of the metal electrodes are of course constant. However, if the activities of the cations Zn^{2+} and Cu^{2+} are unequal, then the e.m.f. will vary from the value E° attained when both are in their standard states. However, if we put the temperature T equal to 298 K, the value of RT/F

is merely 0.02569 V, so that this variation, although real, is modest in its extent except when very large factors are involved in the ratio of the concentrations.

10.3 Half-cells and standard reduction potentials

A cell such as that designated in Equation (10.1) is readily divisible into two. One half is metallic zinc dipping into a solution of Zn^{2+} ions, the other is metallic copper dipping into a solution of Cu^{2+} ions. One cannot measure the e.m.f. of a half-cell on its own: for that one needs a complete cell. What may be done is to use one specific half-cell as an arbitrary standard and measure others against this. The half-cell adopted for this purpose is the standard hydrogen electrode, consisting of an aqueous solution of H^+ of unit activity, through which H_2 at a pressure of 1 bar is bubbling, and into which is dipped a piece of metallic platinum which serves as the electrode.

To measure the e.m.f. of the half-cell, $Zn^{2+}|Zn$, one employs it as the right-hand half of a cell, with the standard hydrogen electrode as the left-hand half, that is:

$$Pt \mid H_2(1 \text{ bar}) \mid H_{aq}^+(1 \text{ mol dm}^{-3}) \parallel Zn_{aq}^{2+} \mid Zn \tag{10.9}$$

Where the Zn^{2+} solution in the right-hand half has unit activity, the measured value is the standard potential, $E°$, for this half-cell. Some standard potentials are listed in Table 10.1. It may be added, as an official health warning, that it is not appropriate to measure all of these in a manner totally analogous to that outlined above. One must not dip a piece of metallic potassium into water and expect it to remain there while one makes careful measurement of its e.m.f. The datum for $K^+|K$ is obtained by an indirect method.

A special type of cell that may be assembled is the concentration cell, which is comprised of half-cells that are identical except that their electrolyte solutions are of different concentrations.

As an example, we may cite the cell,

$$Zn \mid Zn^{2+}(c_1) \parallel Zn^{2+}(c_2) \mid Zn \tag{10.10}$$

for which the cell reaction is:

$$Zn^{2+}(c_2) \quad = \quad Zn^{2+}(c_1) \tag{10.11}$$

In this case, the Nernst equation is simplified because $E°$ is zero. We then have:

$$E = \frac{RT}{2F} \ln \frac{c_1 \gamma_1}{c_2 \gamma_2} \tag{10.12}$$

where the product $c\gamma$ gives the activity of the solute in each solution. Equation (10.12) illustrates that the sign of the e.m.f. of a concentration cell is totally dependent on which electrolyte concentration is the greater.

One important aspect of concentration cells is that they can serve to document the non-ideality of electrolytic solutions. A measurement of E, combined

Table 10.1   A selection of standard reduction potentials, relating to aqueous solutions at 298 K

Half-cell	Half-cell reaction	E°/V
$Li^+\mid Li$	$Li^+ + e^- = Li$	-3.04
$K^+\mid K$	$K^+ + e^- = K$	-2.92
$Ca^{2+}\mid Ca$	$Ca^{2+} + 2e^- = Ca$	-2.87
$Na^+\mid Na$	$Na^+ + e^- = Na$	-2.71
$Mg^{2+}\mid Mg$	$Mg^{2+} + 2e^- = Mg$	-2.36
$Zn^{2+}\mid Zn$	$Zn^{2+} + 2e^- = Zn$	-0.76
$Cr^{3+}\mid Cr$	$Cr^{3+} + 3e^- = Cr$	-0.74
$Fe^{2+}\mid Fe$	$Fe^{2+} + 2e^- = Fe$	-0.44
$Ni^{2+}\mid Ni$	$Ni^{2+} + 2e^- = Ni$	-0.25
$Sn^{2+}\mid Sn$	$Sn^{2+} + 2e^- = Sn$	-0.14
$H^+\mid H_2 \mid Pt$	$2H^+ + 2e^- = H_2$	0.00
$Cl^-\mid AgCl \mid Ag$	$AgCl + e^- = Ag + Cl^-$	0.22
$Cl^-\mid Hg_2Cl_2 \mid Hg$	$Hg_2Cl_2 + 2e^- = 2Hg + 2Cl^-$	0.33
$Cu^{2+}\mid Cu$	$Cu^{2+} + 2e^- = Cu$	0.34
$Hg_2^{2+}\mid Hg$	$Hg_2^{2+} + 2e^- = 2Hg$	0.79
$Ag^+\mid Ag$	$Ag^+ + e^- = Ag$	0.80
$Hg^{2+}\mid Hg$	$Hg^{2+} + 2e^- = Hg$	0.85
$Pt^{2+}\mid Pt$	$Pt^{2+} + 2e^- = Pt$	1.20
$Au^+\mid Au$	$Au^+ + e^- = Au$	1.68

with the known concentrations of the two solutions, permits the evaluation of the ratio of the two activity coefficients, γ_1/γ_2. Where one solution is quite dilute and the other is concentrated, this ratio is found to deviate significantly from unity.

In section 10.1, we referred to the chemical reactions which take place as positive electricity passes through a cell from left to right. It is appropriate to identify the corresponding processes involved in a right-hand half-cell, and these are listed in Table 10.1.

Two terms which are widely used in chemistry are **oxidation** and **reduction**. To illustrate them let us conceive of a divalent metal, M, reacting with oxygen to yield the oxide, MO, which is then treated with water to give the corresponding hydroxide, $M(OH)_2$. If we regard this as an ionic compound, the original atom, M, has had electrons removed to yield the cation, M^{2+}. This has been brought about by its reaction with oxygen so the process is categorised as an oxidation. The reverse process, whether the treatment of the metal oxide with hydrogen to recover the metal, or the addition of electrons to the M^{2+} ion, is termed reduction.

Clearly, for electrons passing from a metal to a solution of its cation, the process involved is a reduction. For this reason, the E° values listed in the table are frequently referred to as standard reduction potentials.

10.4 Analytical applications of cells

Since the e.m.f. of a cell is a function of the activities of the ionic solutes participating in the cell reaction, it is feasible to use a cell to measure the concentration of one of these species, thus thrusting the Nernst equation into an analytical role.

One species whose estimation is frequently desirable is the ion H^+. It participates in the reaction of the hydrogen electrode, so that it would conceivably be possible to put the solution to be sampled into this electrode and determine the H^+ activity using a concentration cell, with a standard hydrogen electrode as the other half of the cell. This, however, would result in a rather clumsy and dangerous device, since it would be necessary to have an accompanying cylinder of hydrogen, with all the explosion hazards that attend the bubbling of hydrogen through the solutions.

The device widely used in the estimation of H^+, called the pH meter, consists of two half-cells, of which one, sensitive to the H^+ concentration of the solution into which it is dipped, is called the glass electrode. An illustration of this is shown in Figure 10.2. It consists of a sealed tube, of which the bottom part is composed of a thin glass membrane, containing an electrode of silver–silver chloride immersed in an aqueous solution of 0.1 mol dm^{-3} HCl. The potential of this electrode varies linearly with the pH of the solution into which it is dipped.

To complete the cell, a reference electrode is required and this role is usually played by a calomel electrode, calomel being a very old name for mercury(I) chloride, Hg_2Cl_2. In this half-cell, mercury is in contact with Hg_2Cl_2 that is in contact with an aqueous solution of KCl, which is connected by a porous

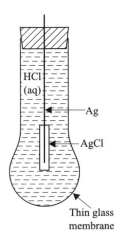

Thin glass
membrane

Figure 10.2 An illustration of the glass electrode, which is the essential component of a pH meter.

liquid junction to the solution whose pH is to be measured. The e.m.f. of such a cell varies with the H^+ activity of the solution into which these electrodes are dipped so that, with careful calibration, the reading obtained can be that of the pH, rather than simply an output in millivolts.

Over the past 30 years, a number of other ion selective electrodes have been developed which enable the concentration of one particular species to be determined electrochemically. A number of such sensors have been developed which can provide important diagnostic information on the state of the seriously ill.

Among simple ions, the appropriate ion selective electrodes are available to provide assay techniques for the measurement of Na^+, K^+, Ca^{2+}, Mg^{2+} and Cl^-. These new electrochemical methods are at least as reliable and much simpler than those used previously, based on flame photometry or colorimetry. The ionic concentrations are all indicative of the state of the patient. For example, low levels of the cation Mg^{2+}, which is linked with several enzymes, are associated with a number of conditions such as hypertension, stroke, seizure and coronary spasm.

Another species whose concentration can be a very good indicator in critically ill patients is the lactate ion, $CH_3CH(OH)CO_2^-$, an intermediate in glucose metabolism. This level is regarded as the best single indicator of clinical shock. In recent years, ion-selective electrodes have been developed to such effect that the direct measurement of lactate ion concentration by electrochemical means is now feasible. This may, conveniently, be performed on a whole blood sample, as distinct from blood serum, which means the results can be available all the more quickly.

While O_2, unlike the species mentioned above, is not an ion, it also may be assayed electrochemically. In the light of Henry's Law, Equation (7.24), it is appropriate to express the O_2 concentration in terms of the partial pressure of O_2 which would be in equilibrium with the concentration actually present, so that blood oxygen levels are often expressed as P_{O_2}. One instrument is available which can determine the pH, P_{O_2} and P_{CO_2} (the partial pressure of CO_2 that would be in equilibrium with the blood CO_2) from a single blood sample. Previously, the last of these parameters was obtainable only by a cumbersome manometric technique. The concurrent measurement of these three parameters provides an important profile of the ventilative state of a patient.

10.5 Redox reactions

For an electrochemical cell consisting of metal electrodes dipping into solutions, as positive electricity moves left to right through the cell, in the right-hand half-cell the electrode will be acting as an electron donor, which means a solute is being reduced. For the left-hand half-cell, the corresponding role of the electrode is as an electron acceptor, and the solute is undergoing oxidation.

Alternatively, it is possible to have a reaction in solution in which both oxidation and reduction are achieved by direct electron transfer. These are called redox reactions. As an example, we may cite the reaction, in acidified aqueous solution, using two transition metals of variable valency,

$$Ce^{4+} + Fe^{2+} = Ce^{3+} + Fe^{3+} \tag{10.13}$$

in which the iron ion is oxidised while the cerium ion is reduced. If one starts off with an acidified solution of Fe^{2+}, to which acidified Ce^{4+} is added, the progress of this redox titration may be followed electrochemically, much as one might monitor the neutralisation of HCl solution by NaOH solution using a pH meter. For this potentiometric titration, the sensitive electrode is a piece of platinised platinum, with a calomel electrode serving as the reference electrode.

The role of the platinum electrode here is that of an electron acceptor or donor for the redox reaction. Initially, the ions of iron predominate and it is appropriate to focus on the reduction process:

$$Fe^{3+} + e^- = Fe^{2+} \tag{10.14}$$

The standard reduction potential will be achieved by the platinum electrode when both of these ions are present at unit activity; however, this will also be achieved when the concentrations of Fe^{3+} and of Fe^{2+} are equal, and the relevant value is $+0.77$ V. For the corresponding reduction of Ce^{4+}, the reduction potential is $+1.61$ V.

In the course of the redox titration, the e.m.f. of the platinum electrode will vary in accordance with the Nernst equation:

$$E = E^\circ - \frac{RT}{F} \ln \frac{[Fe^{2+}]}{[Fe^{3+}]} \tag{10.15}$$

If, initially, a moles of Fe^{2+} are present, then when y moles of Ce^{4+} have been added (where $y < a$), leading to the oxidation of y moles of Fe^{2+} to Fe^{3+}, the potential becomes:

$$E = E^\circ - \frac{RT}{F} \ln \left(\frac{a-y}{y} \right) \tag{10.16}$$

This implies a very rapid rise in E at the very beginning of the titration, since the initial concentration of Fe^{3+} was negligible. The most gradual rise in E will be at $y = a/2$, where there will be a point of inflexion. As y approaches a, there will again be a very sharp rise in E, as the Fe^{2+} ion is being depleted, as illustrated in Figure 10.3, and the stage of steepest ascent of the e.m.f. marks the end-point of the titration.

Beyond this point, it is more convenient to consider the potential in terms of the ions of cerium. When the total amount of Ce^{4+} added has attained $2a$, and half has participated in Reaction (10.13), the concentrations of Ce^{4+} and Ce^{3+} will be equal. The potential of the platinum electrode will then be that of the standard reduction potential of Ce^{4+}/Ce^{3+}, which is $+1.61$ V. In this regime where $y > a$, the relevant equation for the potential is now:

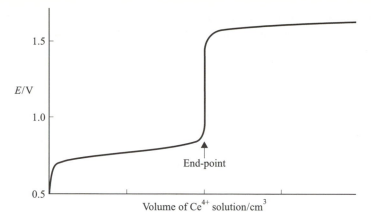

Figure 10.3 A plot of E against the volume of Ce^{4+} solution added to an acidified solution containing Fe^{2+} ion, showing the end-point at the stage of greatest slope.

$$E = E^\circ - \frac{RT}{F}\ln\left(\frac{a}{y-a}\right) \qquad (10.17)$$

Standard reduction potentials readily serve as a guide to indicate which redox reactions are feasible. As applied to Reaction (10.13), they show that the reduction potential of the oxidising agent (Ce^{4+}) exceeds that of the oxidised form (Fe^{3+}) of the reducing agent. On that basis, the reaction is feasible on thermodynamic grounds, but it is impossible to guess how rapidly it might occur. This conclusion may be linked to our discussion of the thermodynamics of chemical equilibrium in Chapter 8. We may evaluate $\Delta G^{\ominus}$ for Reaction (10.13) as $-nF(E_1^\circ - E_2^\circ)$, where E_1° is the standard reduction potential of Ce^{4+}/Ce^{3+} and E_2° is the standard reduction potential of Fe^{3+}/Fe^{2+}, and this leads to the answer, -81.0 kJ mol^{-1}. This means that at 298 K the equilibrium constant K_c for Reaction (10.13) is 1.6×10^{14}, so that if equimolar amounts of Ce^{4+} and Fe^{2+} are mixed, the reaction goes more than 99.99% of the way to completion.

10.6 Bioenergetics

During the Industrial Revolution, the motive power to drive machinery was generated from devices such as the steam engine. In this, a fuel is burned in air and as much as possible of the enthalpy of combustion which is thereby released is employed to boil water to generate steam. The fraction of the heat passed into the steam which can subsequently be converted into work is of course limited by the Second Law of Thermodynamics, so that overall only a modest fraction of the enthalpy of combustion is harnessed for industrial purposes.

179

At a later stage, when space travel was being planned, there developed a need for a means of converting chemical energy into electrical energy within a spacecraft. The solution adopted for the Apollo missions to the moon was to use a hydrogen/oxygen fuel cell, in which the overall chemical reaction was:

$$2H_2(g) + O_2(g) \quad = \quad 2H_2O(l) \qquad\qquad (10.18)$$

The standard Gibbs energy change for this reaction is -474.2 kJ mol^{-1} and in a fuel cell there is the potential to extract an amount of electrical work equal to the extent of the release in the Gibbs energy.

In the processes within the fuel cell by which Reaction (10.18) is achieved, four electrons are transferred. Dividing $\Delta G^{\ominus}$ by $-nF$ yields the theoretical value of 1.229 V for the e.m.f. of the fuel cell: in practice, a figure of 1.2 V is achieved when a moderate current is drawn from it. Clearly, the technology of the late twentieth century can attain a higher efficiency than that of the late eighteenth century.

In an animal, food is metabolised to provide warmth and also a source of energy for doing physical work. The manner in which the second of these is achieved bears little resemblance to the workings of a steam engine. The bodies of animals maintain a temperature close to 37°C, with small temperature gradients. On the other hand, the metabolic processes bear some resemblances to the functioning of a fuel cell. For the latter, a special catalyst had to be developed for the controlled reaction of hydrogen with oxygen: the body contains many specialised catalysts in the form of enzymes. Also, as in a fuel cell, an animal body retains the source of energy in a chemical form until it is required. This storage is achieved chiefly in the form of special deposits of carbohydrates in the muscles and the liver.

When energy is required, hydrolysis of some of the deposits of glycogen generates glucose, which undergoes eventual oxidation to carbon dioxide and water. In the course of the various stages of this intracellular oxidation, the Gibbs energy released is packaged in a special form, creating compounds such as adenosine triphosphate (ATP), which may be used for a variety of purposes, ranging from the overtly physical to the more chemical. These include the contraction of muscles, the operation of the 'pumps' which maintain a larger concentration of sodium ions inside than outside the cells of the body and the synthesis of proteins.

10.7 The biochemical role of adenosine triphosphate

The nucleoside adenosine consists of a bicyclic base called adenine covalently bonded to a C_5 sugar, ribose. This may react with a molecule of phosphoric acid, H_3PO_4, which at the pH of the relevant biological fluids exists partly as $H_2PO_4^-$ and partly as HPO_4^{2-}, to form adenosine monophosphate (AMP), in which the phosphate group is attached to the ribose residue. In the same way, with the elimination of a molecule of water, another molecule of phosphoric acid may react with AMP to yield adenosine diphosphate (ADP). In a repetition of this process, ADP may be converted into adenosine triphosphate (ATP):

$$\text{Adenine—ribose} - \text{O} - \overset{\displaystyle \overset{\text{O}}{\|}}{\underset{\underset{\text{O}^-}{|}}{\text{P}}} - \text{O} - \overset{\displaystyle \overset{\text{O}}{\|}}{\underset{\underset{\text{O}^-}{|}}{\text{P}}} - \text{O} - \overset{\displaystyle \overset{\text{O}}{\|}}{\underset{\underset{\text{O}^-}{|}}{\text{P}}} - \text{O}^-$$

This compound has unusual properties in that the standard Gibbs energy change for the hydrolysis reaction which removes the third molecule of phosphoric acid,

$$\text{ATP} + \text{H}_2\text{O} \quad = \quad \text{ADP} + \text{H}_2\text{PO}_4^- \tag{10.19}$$

is negative to the extent of approximately -30 kJ mol^{-1}. This means that ATP has a strong tendency to undergo hydrolysis: however, this process requires the appropriate enzyme to bring it about.

In the metabolism of glucose, a most significant intermediate is glucose-6-phosphate, which can be formed by the reaction of glucose with phosphoric acid. However, for the reaction,

$$\text{Glucose} + \text{H}_2\text{PO}_4^- \quad = \quad \text{Glucose-6-phosphate} + \text{H}_2\text{O} \tag{10.20}$$

the standard Gibbs energy change is approximately $+13 \text{ kJ mol}^{-1}$, which means that it can occur only to a negligible extent before equilibrium is achieved. However, if Reactions (10.20) and (10.19) were to be coupled,

$$\text{ATP} + \text{Glucose} \quad = \quad \text{ADP} + \text{Glucose-6-phosphate} \tag{10.21}$$

then we have a reaction for which $\Delta G^{\ominus}$ is approximately -17 kJ mol^{-1}. The substantially negative value means that at equilibrium, Reaction (10.21) lies well to the right, going virtually to completion.

In the metabolic pathway, glucose-6-phosphate isomerises to fructose-6-phosphate (where fructose is another C_6 sugar molecule, isomeric with glucose). The next significant step is the attachment of another molecule of phosphoric acid to this molecule, to facilitate its metabolic oxidation into two C_3 compounds, both of which are phosphates. However, for the reaction,

$$\text{Fructose-6-phosphate} + \text{H}_2\text{PO}_4^- = \text{Fructose-1,6-bisphosphate} + \text{H}_2\text{O} \tag{10.22}$$

the standard Gibbs energy change is approximately $+16 \text{ kJ mol}^{-1}$, which must preclude its occurrence on its own. When it is coupled with Reaction (10.19), we obtain the reaction,

$$\text{ATP} + \text{Fructose-6-phosphate} \quad = \quad \text{Fructose-1,6-bisphosphate} + \text{H}_2\text{O} \tag{10.23}$$

for which $\Delta G^{\ominus}$ is approximately -14 kJ mol^{-1}. Clearly, this coupled reaction also may progress virtually to completion.

In later reactions along the metabolic pathway, steps in the metabolism of the C_3 units which have significantly negative standard Gibbs energy changes are coupled to the reverse of Reaction (10.19), thus regenerating ATP. This key intermediate may then serve to expedite reactions which are otherwise impossible. These reactions, achievable with the help of ATP, are not restricted to processes in which a molecule of phosphoric acid is being attached to another intermediate. They may involve the transfer of a totally different group, with the overriding proviso that the Gibbs energy change for the coupled reaction is favourable.

Suggested reading

VANATA, J., 1989, *Principles of Chemical Sensors*, New York: Plenum Press.

YOUNG, C. C., 1997, Evolution of blood chemistry analyzers based on ion selective electrodes, *Journal of Chemical Education*, **74**, 177–82.

KLOTZ, I. M., 1978, *Energy Changes in Biochemical Reactions*, New York: Academic Press.

WRIGGLESWORTH, J. M., 1997, *Energy and Life*, London: Taylor & Francis.

Problems

10.1 Given the standard electrode potentials in Table 10.1, evaluate the e.m.f. of the cell:

$$Cu \mid Cu_{aq}^{2+} \text{ (0.50 mol dm}^{-3}) \parallel Zn_{aq}^{2+} \text{ (0.02 mol dm}^{-3}) \mid Zn$$

10.2 Given the standard electrode potentials in Table 10.1, evaluate the standard Gibbs energy change for the reaction:

$$Ca(s) + 2Ag^+(aq) = Ca_{(aq)}^{2+} + 2Ag(s)$$

10.3 On the basis of the information in Table 10.1, which of the following metals will not be capable of displacing hydrogen from water or from dilute aqueous HCl:

Zn, Ca, Ag, Mg, Li, Hg, Fe, Cu.

Give your reasons.

10.4 Given the following standard reduction potentials:

Reduction process		E^θ/V
$H_2O_2 + 2H^+ + 2e^-$	$= 2H_2O$	1.77
$Br_2 + 2e^-$	$= 2Br^-$	1.06
$Fe^{3+} + e^-$	$= Fe^{2+}$	0.77
$I_2 + 2e^-$	$= 2I^-$	0.54
$Sn^{4+} + 2e^-$	$= Sn^{2+}$	0.15

(a) Is it feasible for Fe^{3+} to oxidise Sn^{2+}?

(b) If Br_2 is added to an aqueous solution of KI, what reaction will take place?

(c) In acid solution, can hydrogen peroxide oxidise bromide ion?

(d) Evaluate the equilibrium constant of the reaction,

$$2Fe^{3+} + 2I^- = 2Fe^{2+} + I_2$$

10.5   The alkali metal ions Na^+ and K^+ are both present *in vivo*, with the sodium ions predominantly extracellular whereas the potassium ions are mostly located inside the cell. Typical concentrations of K^+ ions are 5×10^{-3} mol dm^{-3} outside the cell and 130×10^{-3} mol dm^{-3} within it. Evaluate the consequent potential difference across the cell wall.

11

Chemical kinetics

In considering the rapidity of the progress of a chemical reaction towards equilibrium, we look at experimental techniques by which one may measure the extent of reaction that has occurred, at well-defined points in time, at a particular temperature.

To interpret these data, chemists employ an empirical framework which uses the defined terms: **reaction rate**, **reaction order** and **rate constant**. For the influence of temperature on chemical reactions, it is found that the variation of the rate constant is usually well described by the Arrhenius equation, which features two arbitrary constants.

The rationale of the various equations referred to above is not in any way based on any presuppositions as to how the particular reaction under study is thought to occur: rather, it rests on the fact that these equations serve to describe the kinetic behaviour of the reaction.

11.1 Monitoring the progress of a chemical reaction

Where a well-defined chemical reaction is known to take place, it is usually of interest to document the rate at which the reaction occurs under defined conditions. For present purposes, we shall focus our attention primarily on reactions in solution. Two important parameters then are the identity of the solvent and the temperature.

The experimental approach most used in the early part of this century was to remove aliquots from the reaction mixture at known times, arrest the reaction and then subject the sample to chemical analysis, either for a reactant or for a reaction product. For example, for the reaction, occurring under reflux, of chloroacetate ion with hydroxide ion,

$$\text{ClCH}_2\text{CO}_2^- + \text{OH}^- \quad = \quad \text{HOCH}_2\text{CO}_2^- + \text{Cl}^- \tag{11.1}$$

the reaction may be arrested by delivering the 5 cm^3 sample into 20 cm^3 of ice/water and the concentration of OH$^-$ ion may be assayed by titrating with standard HCl solution. In this way, the OH$^-$ concentration can be measured at frequent intervals as the reaction proceeds. The initial concentrations of both the reactant species would have been known, and since they react together in the ratio of 1:1, from the known concentration of OH$^-$ ion, that of the chloroacetate ion at the same point in time can readily be deduced.

In more recent times, it has become more usual to monitor the progress of a chemical reaction by the continual measurement of some physical property of the reaction mixture. In this way the occasional disturbance to the reaction system caused by the periodic removal of samples is avoided. The range of possible physical properties that may be employed here is quite considerable.

As an example of this approach, let us consider the alkaline hydrolysis of an ester:

$$CH_3CO_2C_2H_5 + OH^- \quad = \quad CH_3CO_2^- + C_2H_5OH \tag{11.2}$$

It is feasible to evaluate the extent of the reaction by monitoring the electrical conductivity of the solution. Initially, the only contributions to the specific conductance, κ, of the solution are from the OH$^-$ ion and its counterion, probably Na$^+$ or K$^+$. As the reaction proceeds, OH$^-$ is replaced progressively by ethanoate anion, which has an appreciably lower conductance, and κ falls in a manner reflecting the progress of the reaction.

However, the technique which has been most widely used in this regard is UV spectrophotometry. In many cases, a spectrophotometer cell will serve as the reaction vessel and, since it is important to control the temperature, an appropriate thermostatic device is needed so that the temperature of the sample cell can be held constant in the cell compartment of the spectrophotometer. Ideally, either a reactant or a product of the reaction should have a fairly strong absorption within the wavelength range of the instrument, so that, at the λ_{max} value of this species, the value of the absorbance, A, is proportional to the concentration of one of the participating species in the reaction.

For example, when iodide ion is oxidised in an acidic solution of hydrogen peroxide,

$$H_2O_2 + 2H^+ + 3I^- \quad = \quad 2H_2O + I_3^- \tag{11.3}$$

the reaction can readily be followed by observing the increase in the absorbance at 350 nm, due to the I$_3^-$ ion. None of the other species present absorbs at this wavelength. However, in interpreting the results, one should also be aware of Equilibrium (8.11) involving the species I$^-$, I$_2$ and I$_3^-$: if the actual concentration of iodide ion is less than approximately 0.3 mol dm^{-3}, then allowance should be made for the product present as I$_2$, which will not be detected at 350 nm.

The progress of some reactions may be followed using what are called 'clock' methods. In these, a small quantity of a compound that readily reacts with one of the products is added to one of the reactants before the reaction is started, along with another compound which effects a change in colour when that product is present. However, it is essential that the presence of these two species causes no interference with the reaction whose kinetics are being studied.

Reaction (11.3) may be studied by a 'clock' method, in which the solutions, after thorough mixing, are watched carefully so that the interval of time until the first appearance of a blue coloration may be noted. To achieve this, before the KI solution has been added to that of H_2O_2, one should add a small amount of sodium thiosulfate along with two drops of a starch solution. The thiosulfate will react quantitatively and rapidly with iodine, as it is produced from reaction (11.3):

$$I_3^- + 2S_2O_3^{2-} = 3I^- + S_4O_6^{2-} \tag{11.4}$$

Until this small amount of $Na_2S_2O_3$ has been consumed, no iodine is permitted to accumulate. Once it has been used up, the blue colour of starch/iodine will be seen to appear, signalling that a calculable extent of Reaction (11.3) has been achieved.

11.2 Reaction rate and reaction order

In order to use kinetic data obtained by means such as those described in section 11.1, it is essential first to define the term **reaction rate**. For a reaction occurring in a homogeneous phase, it denotes the rate of change of concentration as a consequence of the reaction. This may be either the rate of decrease of concentration of a reactant, on the left-hand side of the stoichiometric equation, or the rate of increase of the concentration of a product, on the right-hand side.

As applied to Reaction (11.3), we are immediately aware of a problem: while the rate of decrease of the concentration of H_2O_2 will be equal to the rate of increase of the concentration of I_3^- ion, neither will be the same as the rate of decrease of the concentration of H^+ nor the rate of decrease of that of I^-. Clearly, in this matter the stoichiometric factors cannot be ignored. Secondly, we must recognise that the only exact way to specify rates of change is to use the terminology of calculus.

So, to be more precise, the rate of Reaction (11.3) may be written as any of the following:

$$\text{rate} = -\frac{d[H_2O_2]}{dt} = -\frac{1}{2}\frac{d[H^+]}{dt} = -\frac{1}{3}\frac{d[I^-]}{dt} = \frac{d[I_3^-]}{dt} \tag{11.5}$$

In principle, the value of any one of these at some point in time can be determined by measuring the slope of the tangent to the curve, at that point, of the concentration of the relevant species as a function of time.

It is found, empirically, that the rate of a chemical reaction usually shows a proportionality to the concentration of each reactant to a certain power. As applied to Reaction (11.3), for example, the following proportionality is anticipated,

$$-\frac{d[H_2O_2]}{dt} \propto [H_2O_2]^p[H^+]^q[I^-]^r \tag{11.6}$$

where p is called the order of the reaction with respect to H_2O_2, q is called the order of the reaction with respect to H^+, and r is called the order of the reaction with respect to I^-. There is no means of knowing, in advance of their determination by experiment, what are the values of p or q or r, or of the sum $(p + q + r)$, which is usually called the overall order of the reaction.

Where a proportionality exists, an equation may be obtained by inserting the appropriate constant of proportionality. In kinetic equations, this is written as k (lower case) and is called the rate constant. Thus we have:

$$-\frac{d[H_2O_2]}{dt} = k[H_2O_2]^p[H^+]^q[I^-]^r \tag{11.7}$$

The rate constant, k, is equal to the numerical value of the rate when the concentrations of all reactants are one. The units of k are of course dependent on the value of the overall reaction order. The left-hand side of Equation (11.7) has the dimensions, concentration $\times$ (time)$^{-1}$. On the right, after k we have (concentration)^n, where n is the overall order. Consequently, k must have the dimensions, (concentration)$^{n-1}$ (time)$^{-1}$. So if $n = 1$, i.e. the reaction is first order, k will have the units, s^{-1}. If $n = 2$ and the reaction is second order, k will have the units, dm^3 mol^{-1} s^{-1}.

Before developing these ideas, it may be helpful to offer an explanatory word. In the major areas of physical chemistry, such as thermodynamics or spectroscopy, we deal with quantities that can carefully be defined and are expressed in equations that can be derived mathematically. Just occasionally, within this theoretical framework, there are quantities of a much more empirical nature, capable of exact definition but determined only by experiment. In thermodynamics, the activity coefficient is one of these empirical quantities.

As an area of physical chemistry, chemical kinetics differs from these others in that it has a complete empirical framework, used to rationalise the results of experimental studies and to report the kinetics of reactions in an intelligible and useful form. In pursuing that theme, we leave to one side the understandable curiosity as to how the reactions we are studying may take place. However, these questions are merely being postponed until Chapter 12.

11.3 Integrated rate equations: first order reactions

A rate equation necessarily involves differential quantities, which refer to the slope of a particular curve of concentration against time. By the process of integration (see Appendix Two) such an equation can be modified into one that interrelates concentrations with time and is rather more easily employed.

Suppose we have the reaction,

$$A = L \tag{11.8}$$

occurring in aqueous solution, and that the initial concentration of A at $t = 0$, is given by a. It is convenient to use x as a variable in regard to the changing concentrations and to say that at time t, the concentration of A has fallen to $(a - x)$, with that of L rising to x. So for the rate of this reaction, at any point in time, we may write:

$$\text{rate} = -\frac{d(a - x)}{dt} = \frac{dx}{dt} \tag{11.9}$$

If this reaction is first order, as it may be, then we have,

$$\frac{dx}{dt} = k_1(a - x) \tag{11.10}$$

where k_1 denotes the appropriate first order rate constant. To integrate, we first separate the variables, x and t:

$$\frac{dx}{(a - x)} = k_1 dt \tag{11.11}$$

Integrating each side we have:

$$-\ln(a - x) = k_1 t + \text{const} \tag{11.12}$$

To evaluate the constant of integration, we use the fact that at $t = 0$, $x = 0$. It follows that the constant is $-\ln a$, so we have,

$$\ln a - \ln(a - x) = k_1 t \tag{11.13}$$

$$\text{or } \ln\left(\frac{a}{a - x}\right) = k_1 t \tag{11.13a}$$

$$\text{or } (a - x) = ae^{-k_1 t} \tag{11.13b}$$

So, if a reaction follows first order kinetics, it follows from Equation (11.13) that the plot of the natural logarithm of the reactant concentration against time is linear, with a slope of $-k_1$. If we plot $\ln\{a/(a-x)\}$, then the line passes through the origin with a slope of k_1. Equation (11.13b) shows that the curve of reactant concentration against time will be an exponential decay.

From Equation (11.13a) we can readily obtain an expression for the half-life, $t_{1/2}$, the interval required for the reaction to proceed halfway to completion. Putting $x = a/2$, we have:

$$\ln\left(\frac{a}{a/2}\right) = \ln 2 = k_1 t_{1/2} \tag{11.14}$$

$$\therefore t_{1/2} = \frac{\ln 2}{k_1} = \frac{0.693}{k_1} \tag{11.15}$$

189

This means that $t_{1/2}$ is a function only of the rate constant, k_1, and is independent of a, the initial concentration. We referred to this in connection with radioactive decay, a first order process, in Chapter 1.

We want now to look at how one uses experimental kinetic data to deduce the reaction order and to evaluate the relevant rate constant.

— EXAMPLE 11.1

In a chemical reaction, the concentration of the reactant, A, was found to vary with time as follows:

$[A]/(10^{-3}\ \text{mol dm}^{-3})$	41.6	29.8	15.6	11.8	5.8
t/s	0	100	292	376	590

Show that the reaction follows first order kinetics and evaluate the rate constant. Also, determine the half-life of this reaction.

— SOLUTION

At any value of t, the ratio, $a/(a-x)$ is the initial concentration of A divided by the concentration at time t. So it is a simple matter, with a calculator, to evaluate $\ln\{a/(a-x)\}$.

$\ln\left(\frac{a}{a-x}\right)$	0	0.334	0.981	1.260	1.970

The plot of this function against time is shown in Figure 11.1 and gives a good straight line. From the slope we have the rate constant, $k = 3.35 \times 10^{-3}\ \text{s}^{-1}$.

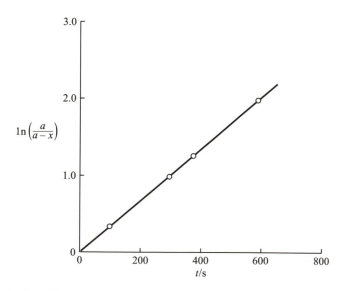

Figure 11.1 A plot of $\ln\{a/(a-x)\}$ against t, using the data of Example 11.1.

Alternatively, by rearranging Equation (11.13a), we can divide $\ln\{a/(a-x)\}$ by t and evaluate k from each experimental point after zero time. This gives:

k/s^{-1}	–	0.00334	0.00336	0.00335	0.00334

The fact that these calculated values of k are constant means that Figure 11.1 must be linear. In this era of the pocket calculator, this is arguably a faster way of testing the applicability of Equation (11.13a) than plotting the graph.

The half-life is most easily evaluated from Equation (11.15):

$$t_{1/2} = \frac{\ln 2}{3.35 \times 10^{-3} \mathrm{s}^{-1}} \tag{11.16}$$
$$= 206.9\mathrm{s}$$

— EXAMPLE 11.2 —

The hydrolysis of methyl bromide in aqueous solution,

$$CH_3Br + H_2O = CH_3OH + H^+ + Br^- \tag{11.17}$$

was monitored by titrating aliquots of the reaction mixture with $AgNO_3$ solution. This enables the concentration of Br^- ion to be determined. In an experiment at 330 K, the following volumes, v, were required to precipitate the Br^- ion in aliquots of 10 cm^3.

t/min	0	88	196	300	412	540	∞
v/cm^3	0	5.9	12.1	17.3	22.1	26.7	49.5

Show that the solvolysis follows first order kinetics and evaluate the rate constant.

— SOLUTION —

It is important to appreciate that v is a measure of the concentration of Br^-, which is a *product* of this reaction. In terms of the variables in Equation (11.13a), v is a measure of x. The infinity value of x, at 49.5, is a comparable measure of a. Thus the ratio $a/(a-x)$, is given by $49.5/(49.5-v)$. On that basis we can evaluate $\ln\{a/(a-x)\}$, even though we have not been told the concentration of the $AgNO_3$ solution.

t/min	0	88	196	300	412	540
$\ln\left(\frac{a}{a-x}\right)$	0	0.127	0.280	0.430	0.591	0.775
$\frac{1}{t}\ln\left(\frac{a}{a-x}\right)$	–	0.00144	0.00143	0.00143	0.00144	0.00144

Clearly, from the constancy of the figures on this last line, the reaction is first order and the rate constant $k_1 = 1.44 \times 10^{-3}$ min^{-1}. Equally, plotting $\ln\{a/(a-x)\}$ against t will lead to a straight line graph, whose slope will give the same answer.

11.4 Integrated rate equations: second order reactions

If we now consider the reaction,

$$A + B \quad = \quad L + M \tag{11.18}$$

assume that it is first order with respect to each reactant and designate the concentrations as follows:

	A	B	L	M
$t = 0$	a	b	0	0
$t = t$	$(a - x)$	$(b - x)$	x	x

For the rate equation, we then have,

$$\frac{dx}{dt} = k_2(a - x)(b - x) \tag{11.19}$$

where k_2 denotes the appropriate second order rate constant. We may rewrite this:

$$\frac{dx}{(a - x)(b - x)} = k_2 dt \tag{11.20}$$

The integration of the left-hand side requires us to specify whether a and b are equal or unequal. First, let us deal with the case where $a = b$: that is, we have equal concentrations of the two reactants initially and since they react together in the ratio of one molecule of A to one of B, these concentrations will remain equal to each other. This gives:

$$\frac{dx}{(a - x)^2} = k_2 dt \tag{11.20a}$$

Integrating this in the same way as Equation (11.11), we obtain,

$$\frac{1}{(a - x)} - \frac{1}{a} = k_2 t \tag{11.21}$$

as the integrated second order equation. It follows from this that if Reaction (11.18) follows second order kinetics, then the reciprocal of the concentration of A should be a linear function of t, with the slope equal to the second order rate constant, k_2.

At this point, it is fitting to look at how the reaction order affects the way in which the reaction rate slows down as a reaction progresses. In Figure 11.2 we start with a certain reaction rate and look at the different ways in which the reactant concentration falls off if the reaction is (a) zero order or (b) first order or (c) second order. The first of these is a straight line, for the rate is independent of the reactant concentration. One consequence of this is that the half-life is proportional to the initial concentration. For the first order decay, we have an exponential decay as described by Equation (11.13b), with the half-life independent of the initial concentration.

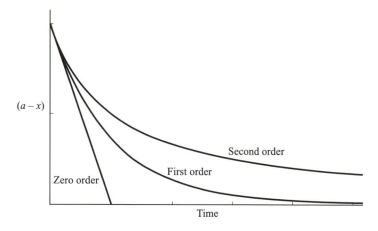

Figure 11.2 Plots of $(a - x)$ against time for reactions of different orders (zero, first and second), all with the same initial concentration and initial rate.

The curve for the second order reaction, based on equation (11.21), represents a much greater slowing down of the reaction as it proceeds. We may evaluate the half-life from Equation (11.21) by substituting $x = a/2$, to obtain:

$$t_{1/2} = \frac{1}{ak_2} \tag{11.22}$$

This demonstrates that the half-life of a second order reaction is inversely proportional to the initial concentration, as may be seen in Figure 11.2.

Alternatively, we may assume that the initial concentrations referred to in Equation (11.19) are different and that $a > b$. In that case the integration of Equation (11.20) is more difficult and it will not be explained here. It yields the equation:

$$\frac{1}{(a - b)} \ln \left\{ \frac{b(a - x)}{a(b - x)} \right\} = k_2 t \tag{11.23}$$

An interesting situation arises if a is very much greater than b, perhaps by a factor in excess of 30. The variable x cannot be greater than b, so that $(a-b)$ and $(a-x)$ are scarcely different from a. If we make those approximations, Equation (11.23) becomes:

$$\left. \begin{aligned} \frac{1}{a} \ln \left(\frac{b}{b - x} \right) &= k_2 t \\ \text{or } \ln \left(\frac{b}{b - x} \right) &= a k_2 t \end{aligned} \right\} \tag{11.24}$$

In this scenario, the concentration of B decreases in accordance with an equation almost identical to Equation (11.13a), relating to a first order reaction. We say that, under these experimental conditions, the second order reaction has become pseudo-first order. In Equation (11.24), the pseudo-first order rate constant, ak_2, is the product of the second order rate constant of the reaction of A and B with the excess concentration of A. So the true rate constant for the reaction of A and B is found by dividing the pseudo-first order rate constant by the excess concentration of A.

—— **EXAMPLE 11.3** ————————————————————————
The progress of the reaction,

$$C_2H_5NO_2 + OH^- \quad = \quad C_2H_5OH + NO_2^- \tag{11.25}$$

was monitored by measuring the hydroxide concentration periodically. In an experiment with the initial concentrations of the reactants $[C_2H_5NO_2]_0 = 0.020$ mol dm^{-3} and $[OH^-]_0 = 0.010$ mol dm^{-3}, the data obtained were as follows:

t/s	30	50	75	100	140	180	240
$[OH^-]/(mol\ dm^{-3})$	0.0071	0.0058	0.0045	0.0037	0.0026	0.0019	0.0013

Show that these data are consistent with the reaction being first order with respect to each reactant and evaluate the rate constant.

—— **SOLUTION** ————————————————————————
The relevant equation here is Equation (11.23) and it is appropriate to let nitroethane represent A and hydroxide ion B. The function, $\ln\{b(a-x)/a(b-x)\}$ is evaluated and in Figure 11.3, it is plotted against t. This gives a good straight line, whose slope is measured and equated to $(a-b)k_2$. We then have:

$$(a - b)k_2 = 0.0063 \text{ s}^{-1}$$
$$\therefore k_2 = 0.63 \text{ s}^{-1}$$

11.5 Determining the reaction order and the rate constant

Where there is only one reactant species involved, the determination of the reaction order is relatively simple. One obtains, by some appropriate experimental method, kinetic data on the progress of the reaction as a function of time and uses these to test whether they fit the appropriate equations for the reaction being first, second or zero order. Admittedly, the order of the reaction is not necessarily an integer, and there are an infinite number of non-integral values between zero and two, but non-integral orders are relatively rare.

Where there are two or more species on the left side of the balanced equation, then one would normally start with experiments involving stoichiometric

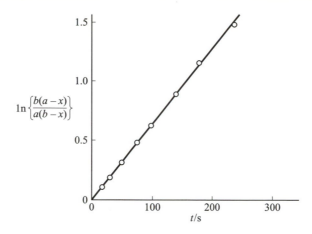

Figure 11.3 A plot of $\ln\{b(a-x)/a(b-x)\}$ against t for the data of Example 11.3.

amounts of each reactant. By the above procedures, one determines the overall order, n, of the reaction. There remains the question as to the individual orders with respect to each reactant species. These can best be approached using the isolation method. If a kinetic experiment is carried out in which every species except one is present in large excess, then the apparent reaction order exhibited during the experiment will be the order with respect to the single reactant that is not present in excess.

When a reaction is being studied under these conditions, it is desirable that the technique used to monitor its progress is sensitive to the concentration of the species not in excess. For example, if it is intensely coloured and is the only such species involved in the reactions, then spectrophotometry is an appropriate method. Whatever the technique, it is important that the change in the parameter being measured is linearly related to the change in the concentration of the reactant species. It follows that if, at $t = 0$, the property has the value χ_0, at $t = t$, it has the value χ_t and when the reaction is complete it has attained χ_∞, then the ratio, $(\chi_0 - \chi_\infty)/(\chi_t - \chi_\infty)$ is equal to $a/(a-x)$. So, if the reaction follows first order or pseudo-first order kinetics, the natural logarithm of this quantity will be a linear function of t. So the relevant rate constant can readily be derived, provided χ_∞ can be measured with some accuracy.

A reaction following first order kinetics will be 99.9% complete after 10 half-lives, so that a reading taken then or later may serve as χ_∞. If, however, the reaction is following second order kinetics, it requires 1000 times the duration of the first half-life before the reaction is 99.9% complete. Such an interval is prohibitively long and in this circumstance χ_∞ is effectively unobtainable.

This is exemplified in the use of flow techniques for the study of reactions in solution. These were first introduced in the 1920s by the biochemists Hartridge and Roughton, who wished to study reactions such as that between

195

haemoglobin and oxygen, which take place over an abbreviated time scale. It was realised that if the two species were dissolved in separate solutions, which were led from their respective storage vessels to be mixed in a flow system, then the distance from the point of mixing would be proportional, for any element of solution, to the interval of time since mixing had occurred. High flow rates would permit good time resolution to be achieved.

Whereas the continuous flow method needed very large volumes of the solutions, the variation which is most widely used now makes no such demands. The stopped-flow technique is illustrated in Figure 11.4. After solutions of the two reactant species are thoroughly mixed, the resulting solution flows along a section of tubing of high optical quality with a rectangular cross-section. The flow of the solutions is then suddenly halted and the absorbance of the mixed solution, at an appropriate wavelength in the UV or visible, is then recorded over an interval of time in the range of tens of milliseconds to tens of seconds, while the reaction progresses. Normally, the concentrations of the two

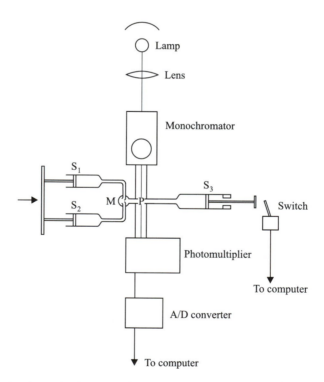

Figure 11.4 A schematic diagram of a stopped-flow instrument, with the reactant solutions in the syringes S_1 and S_2, the mixing chamber denoted by M, the analysing light monitoring the solution at the point P and the mixed solution collected in the syringe S_3.

solutions to be mixed are chosen so that the reaction will occur under pseudo-first order conditions. Then it becomes realistically possible to measure satisfactorily the absorbance value at $t = \infty$. The pseudo-first order rate constant is then determined, for each experiment, from the slope of the plot of $\ln(A_\infty - A_t)$ against time.

In following a reaction by the stopped-flow technique, there are two possible scenarios. One may monitor the appearance of a reaction product or the disappearance of a reactant. (In some instances, it may be possible to do one or the other simply by altering the wavelength of observation.) The resulting traces of absorbance, A, against time for these alternative situations are sketched in Figure 11.5.

In some reaction systems, the products engage, perhaps slowly, in further reactions which may affect the value of the parameter being measured. The effect of this is that A_∞ is distorted to a much greater extent than is A_t at 10–70% reaction, and a reliable value of A_∞ is unobtainable. Despite this, there are means of deducing the rate constant where the reaction is following first order kinetics, but not where it is second order. For these reasons, there is an appreciable advantage in studying a second order reaction under pseudo-first order conditions. The true second order rate constant is then found by taking the slope of the plot of the pseudo-first order rate constant against the concentration of the species present in excess.

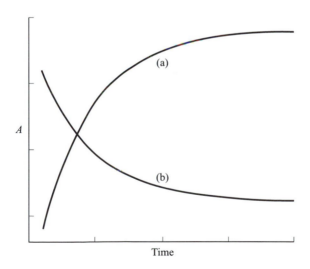

Figure 11.5 Notional alternative results from a stopped-flow experiment: (a) the absorbance rises due to an absorbing reaction product, or (b) the absorbance decreases as the absorbing reactant is used up. In the latter case, this reactant needs to be the one which is not initially present in excess.

11.6 The effect of temperature on reaction rates

In the foregoing discussions it has implicitly been assumed that the temperature is kept constant. We now come to consider what effect a change of temperature has on the rate of a chemical reaction. The first point to make is that its effects can adequately be represented as those on the rate constant.

For most reactions, the effect on the rate constant, k, of varying the temperature is satisfactorily represented by the equation which, in 1889, Arrhenius showed to fit all the data then available to him. This equation is,

$$\ln k = \ln A - \frac{E_a}{RT} \tag{11.26}$$

where T is the temperature on the Kelvin scale, R is the gas constant, the parameter E_a is (for historical reasons) called the activation energy and A is usually called the A factor.

Equation (11.26), usually called the Arrhenius equation, implies that $\ln k$ should be a linear function of T^{-1}. Alternatively, the same relation may be written,

$$k = A\,e^{-E_a/RT} \tag{11.27}$$

which demonstrates that k must increase appreciably as T is increased, in a manner involving an exponential function.

— **EXAMPLE 11.4** —————————————————————————

Given that the rate constant, k, of a decomposition reaction varies with the temperature as follows,

$t/°C$	170	180	190	200
k/s^{-1}	1.92×10^{-4}	4.6×10^{-4}	1.05×10^{-3}	2.34×10^{-3}

evaluate the Arrhenius activation energy of this reaction and the anticipated value of the rate constant at 298 K.

— **SOLUTION** —————————————————————————

We need to evaluate $\ln k$ and T^{-1}, with T in the Kelvin scale.

T/K	443	453	463	473
$T^{-1}/10^{-3}K^{-1}$	2.257	2.207	2.159	2.113
$\ln(k/s^{-1})$	−8.56	−7.68	−6.86	−6.06

In Figure 11.6, $\ln k$ is plotted against T^{-1}, giving an excellent straight line, of which we need to measure the slope.

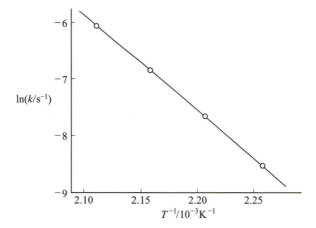

Figure 11.6 A plot of ln k against T^{-1} for the data of Example 11.4.

$$\text{Slope} = \frac{-8.43 - (-6.01)}{(2.25 - 2.11) \times 10^{-3}} \text{K}$$

$$= \frac{-2.42}{0.14 \times 10^{-3}} \text{K}$$

$$= -17.29 \times 10^3 \text{ K}$$

$$E_\text{a} = -R.\text{slope}$$

$$= -8.314 \text{ J K}^{-1} \text{ mol}^{-1} \times -17.29 \times 10^3 \text{ K}$$

$$= 143.7 \text{ kJ mol}^{-1}$$

To obtain A, it is important to note that the scale of the abscissa in Figure 11.6 does not start at zero, so ln A is not available as the intercept. The best way is to substitute the values of k and of T, for a datum point, into Equation (11.26) along with the known value for E_a. Let us substitute at 180°C:

$$\ln A = \ln k + \frac{E_\text{a}}{RT}$$

$$= -7.68 + \frac{143.7 \times 10^3}{8.314 \times 453}$$

$$= -7.68 + 38.15$$

$$= 30.47$$

$$\therefore A = 1.7 \times 10^{13} \text{ s}^{-1}$$

We need this value to calculate the rate constant k, at 298 K.

$$\ln k = 30.47 - \frac{143.7 \times 10^3}{8.314 \times 298}$$

$$= 30.47 - 57.97$$

$$= -27.50$$

$$\therefore k = 1.14 \times 10^{-12} \ s^{-1}$$

(Since this rate constant corresponds to a half-life of more than 19 000 years, one could say that at this temperature the reaction is virtually stationary.)

The Arrhenius equation, although generally followed, is not invariably obeyed in precise detail. Deviations from it come under two main headings. Firstly, there are instances where, despite the accuracy of the experimental data, the plots of ln k against T^{-1} show curvature. To fit such data more exactly, one normally uses an equation with at least three empirical parameters. In other cases, the plot is found to show a sudden change in slope between two linear portions. This usually implies a change, as a function of temperature, in the reaction mechanism and will be alluded to in Chapter 12.

Suggested reading

LOGAN, S. R., 1996, *Fundamentals of Chemical Kinetics*, Harlow: Longman. (Particularly Chapters 1 and 2.)

LAIDLER, K. J., 1987, *Chemical Kinetics*, 3rd Edn, New York: Harper & Row.

HOUSE, J. E., 1997, *Principles of Chemical Kinetics*, Dubuque, IA: Wm C. Brown.

Problems

11.1 For a reaction involving the substances A and B, at constant temperature, the following data were obtained:

[A]/(mol dm^{-3})	[B]/(mol dm^{-3})	Initial rate/(mol dm^{-3} s^{-1})
0.12	0.045	6.5×10^{-4}
0.24	0.090	2.6×10^{-3}
0.72	0.090	7.8×10^{-3}

Assuming that the rate equation is of the form,

Rate $= k[A]^p[B]^q$

determine the orders, p and q, and the rate constant, k.

11.2 In a chemical reaction, the concentration of the reactant A, measured as a function of time, gave the following data:

[A]/(10^{-5} mol dm^{-3})	20.8	14.5	7.1	5.4	2.0
t/s	0	50	150	188	325

Show that the reaction is kinetically first order and evaluate the rate constant. Hence evaluate the half-life of the reaction.

11.3 Benzene diazonium chloride in aqueous solution decomposes:

$$Ph-\overset{+}{N}\equiv N\ Cl^- \quad = \quad Ph-Cl + N_2\uparrow$$

In one experiment the volume of N_2 evolved was measured as a function of time:

t/min	0	116	192	355	481	880	1282	∞
vol of N_2/cm^3	0	10.3	16.2	26.3	32.6	45.7	52.6	60.0

Show that the reaction is first order and evaluate the rate constant.

11.4 In studying the reaction,

$$C_5H_5N + C_2H_5I \quad = \quad C_7H_{10}N^+\ I^-$$

in an experiment at 373 K, with initial concentrations of pyridine and iodoethane both equal to 0.100 mol dm^{-3}, the concentration of I^- ion measured at various times was as follows:

t/s	180	420	710	1080	1500
$[I^-]$/(mol dm^{-3})	0.012	0.024	0.035	0.045	0.053

Show that the reaction follows second order kinetics and evaluate the rate constant.

11.5 For the oxidation of bromide ion by acidic hydrogen peroxide in aqueous solution,

$$2Br^- + H_2O_2 + 2H^+ \quad = \quad Br_2 + 2H_2O$$

the rate equation is

$$v = k[H_2O_2][Br^-][H^+]$$

reflecting the fact that the reaction does not take place in one step.
(a) If the concentration of H_2O_2 were increased by a factor of three, how should the rate of disappearance of Br^- ions be affected?
(b) If, under certain conditions the rate of disappearance of Br^- ions is 7.2×10^{-3} mol dm^{-3} s^{-1}, what should be the rate of disappearance of H_2O_2?
(c) If by the addition of water to the reaction mixture the total volume were doubled, what would be the effect on the rate of disappearance of Br^-?
(d) In the rate expression above, what would be acceptable units for the rate constant, k?

11.6 For a decomposition reaction, the first order rate constant is found to vary with temperature as follows:

$T°$/C	5	15	25	35
k/s^{-1}	1.5×10^{-6}	8.0×10^{-6}	4.1×10^{-5}	2.0×10^{-4}

Construct an Arrhenius plot and thus determine the parameters E_a and A in the equation, $\ln k = \ln A - E_a/RT$.

12

Reaction mechanism and enzymic catalysis

A chemical reaction, involving the rearrangement of the bonding electrons of the reactant molecules, may conceivably come about in a single step. Alternatively, the process may consist of a number of discrete steps which, when aggregated, add up to the same equation. This is the issue of reaction mechanism.

Where a reaction takes place in a single step then the rate will be proportional to the concentration of each participating species. A reaction of more complex mechanism is a little bit like a busy cafeteria: some stages are rapid, others are sluggish. The slow step will tend to be rate determining, and the rate of a multi-step reaction is, in general, independent of the concentration of a reactant which becomes involved only after the slow step.

Chemical mechanisms may be so indirect that a substance not appearing in the balanced equation participates in the reaction mechanism. Such substances are called catalysts. In any living organism, there are special catalysts for a huge range of metabolic and other processes. These substances, all extremely specialised in their function, are called enzymes. By contrast, free radical species tend to react very readily, without the need for any catalysts, but with less specific consequences.

12.1 Reaction mechanism

When a chemical reaction takes place there are two alternative possibilities. It may occur in a single step involving precisely those species on the left-hand side of the balanced equation or it may take place in two or more distinct steps. If the former is true, then the reaction is categorised as an elementary one, whereas in the latter case we speak of it as a reaction with a more complex mechanism.

Reactions that involve more than two chemical species rarely take place in a single step. An elementary reaction involving only one species is described as a unimolecular one, whereas one involving two molecules (or radicals or ions), whether identical or different, is termed a bimolecular one. If the reaction is taking place in solution, the role of the solvent is not always a totally passive one, but the involvement of a molecule of solvent in the reaction is not usually reflected in the molecularity attributed to it.

As an example of a unimolecular process, one could cite any case of radioactive decay listed in Chapter 1. Alternatively, there is the isomerisation of cyclopropane to propene,

$$\underset{\underset{\displaystyle CH_2-CH_2}{\diagup\quad\diagdown}}{CH_2} \longrightarrow CH_2:CH-CH_3 \tag{12.1}$$

or of a substituted cyclopropane to the corresponding non-cyclic product. In each case, the rate under specified conditions is proportional to the number of potentially reactive species present, so a unimolecular reaction will follow first order kinetics.

In this book, balanced equations have been written to indicate the reactants and the products in a chemical reaction. At this point, a new symbol, namely an arrow ($\longrightarrow$), has been introduced, which will be employed only in those cases where the process is an elementary one. The use of the arrow in Equation (12.1) does not merely connect the reactants to the products, it is also an assertion in regard to the mechanism of this reaction. This usage will be observed in the following discussion of reaction mechanism.

A useful illustration of the distinction between the balanced equation for a reaction and its possible mechanism is provided by the hydrolysis of an alkyl halide, RX, in alkaline aqueous solution. The stoichiometric equation is:

$$RX + OH^- \quad = \quad ROH + X^- \tag{12.2}$$

It has been shown that the hydrolysis of an alkyl halide under these conditions may occur by one of two possible mechanisms. In some cases, the reaction is an elementary one,

$$RX + OH^- \quad \rightarrow \quad ROH + X^- \tag{12.3}$$

involving the attack of the OH^- ion on the carbon atom next to the halogen atom, with the halide ion leaving at one side as the hydroxide ion approaches at the other. To this process, physical organic chemists have given the name, bimolecular nucleophilic substitution, and the corresponding label, S_N2. Clearly, the rate of this reaction must be proportional to the concentration of RX and to the concentration of hydroxide ion.

$$v_a = k_2[RX][OH^-] \tag{12.4}$$

For some other alkyl halides, the hydrolysis follows a different mechanism involving two steps. The first of these involves only the alkyl halide, wherein it undergoes heterolytic dissociation:

$$RX \quad \rightarrow \quad R^+ + X^- \tag{12.5}$$

For this to occur, it is necessary that the ion R^+ is fairly stable in solution. The name given to this process, involving only a single molecule of reactant, is a unimolecular nucleophilic substitution, labelled S_N1. It is then followed by the rapid step:

$$R^+ + OH^- \quad \rightarrow \quad ROH \tag{12.6}$$

Clearly, the combination of Reactions (12.5) and (12.6) adds up to Reaction (12.2), just as does the S_N2 process. The rate of the S_N1 process is dictated by the slow initial step, (12.5), for which we have the rate expression:

$$v_b = k_1[RX] \tag{12.7}$$

Thus the rate of the S_N1 reaction is independent of the hydroxide concentration, whereas that of the S_N2 reaction is proportional to it.

For some alkyl halides under certain conditions the hydrolysis of RX may proceed, in parallel, by both an S_N1 and an S_N2 mechanism. The rate of disappearance of RX is then given by the sum of the rates of the two solvolytic processes,

$$-\frac{d[RX]}{dt} = v_b + v_a$$
$$= k_1[RX] + k_2[RX][OH^-]$$
$$= \{k_1 + k_2[OH^-]\}[RX] \tag{12.8}$$
$$= k'[RX]$$

This equation shows that the reaction will be first order with respect to RX, but the value of the effective first order rate constant, k', will depend on the concentration of OH^-. A plot of k' against the OH^- concentration should be linear, with the slope equal to k_2 and the intercept on the ordinate equal to k_1.

There is an additional distinction between S_N1 and S_N2 mechanisms, which may be described in regard to the alkaline hydrolysis of an alkyl halide. Suppose the alkyl group, R, consists of three different substituents attached to the carbon atom to which the halogen atom, X, is bonded, The compound, RX, will then be optically active. If we start off with a single enantiomer of RX, should the solvolysis occur by an S_N1 process, then a carbenium ion R^+ will be formed in which the central carbon atom is sp^2 hybridised, just as is the boron atom in the molecule BF_3 (see p. 45). Thus this carbon atom and the bonds to its three substituents will all lie in the same plane. When a hydroxide ion comes to react with this carbenium ion, approach from either side will be equally likely, so that the resulting alcohol should be a racemate, containing equal amounts of the two enantiomers.

On the other hand, if the hydrolysis occurs by an S_N2 mechanism, no such intermediate is involved. The hydroxide ion attacks the central carbon atom from the side opposite to that occupied by the X atom, and in a concerted process, OH^- approaches from one side as X^- departs from the other.

Consequently, a single enantiomer of the alcohol will be formed. However, optical inversion will have taken place, so that the configuration of the R group will now be the reverse of that in the initial RX molecule.

12.2 The kinetics of a reversible reaction

In Chapter 11, we made the implicit assumption that the standard Gibbs energy change for a reaction which is the subject of kinetic study is so negative that the reaction will go virtually to completion. We wish now to look at the kinetic consequences if this is not the case and the rate of the reverse reaction is not totally negligible, so that an appreciable amount of the reactant remains at equilibrium.

Suppose the process involved is the interconversion of A and B, and that both the forward and reverse reactions are first order, with rate constants k_1 and k_{-1}. Also, let us assume we start with only A present, at a concentration a.

	A	=	B	(12.9)
$t = 0$	a		0	
$t = t$	$(a - x)$		x	
$t = \infty$	$(a - x_e)$		x_e	

We have chosen the parameter, x_e, to denote the equilibrium concentration of B. To evaluate x_e, let us equate the forward and the reverse rates at equilibrium. This gives:

$$k_1(a - x_e) = k_{-1}x_e \tag{12.10}$$

which leads to:

$$x_e = \frac{k_1 a}{k_1 + k_{-1}} \tag{12.11}$$

For the net rate of the reaction, at an earlier stage, we have:

$$-\frac{d[A]}{dt} = \frac{dx}{dt} = k_1(a - x) - k_{-1}x \tag{12.12}$$

In this expression, we may substitute for k_1a from Equation (12.11), to obtain:

$$\frac{dx}{dt} = (k_1 + k_{-1})x_e - k_1x - k_{-1}x$$
$$= (k_1 + k_{-1})(x_e - x) \tag{12.12a}$$

Integrating this equation in the manner described in Chapter 11, this leads eventually to the equation:

$$\ln\left(\frac{x_e}{x_e - x}\right) = (k_1 + k_{-1})t \tag{12.13}$$

This expression closely resembles Equation (11.13a) for a simple first order reaction, but there are two significant differences. Firstly, the parameter whose logarithm is a linear function of time is $(x_e - x)$ which, unlike $(a - x)$ in Equation (11.13), is not the concentration of any species actually present. Secondly, from the straight line which results from plotting $\ln(x_e - x)$ against t, one obtains the sum of the forward and reverse rate constants. The individual values of k_1 and k_{-1} can be determined from this using Equation (12.11).

12.3 Catalysis by an acid

A reaction which exhibits very interesting kinetic behaviour in aqueous solution is the iodination of 2-propanone (acetone) in the presence of an acid. The balanced equation is:

$$CH_3COCH_3 + I_2 \quad = \quad CH_3COCH_2I + HI \qquad (12.14)$$

From experimental study, it is found that the rate of this reaction is proportional to the concentration of 2-propanone, is independent of the concentration of I_2 and is proportional to the concentration of H^+. Clearly, in the light of these findings, Reaction (12.14) is not an elementary process.

Extensive studies of this and related reactions have been made, in the light of which it is thought that the mechanism of this reaction is now well understood. This mechanism will now be presented and it will be demonstrated that it carries the kinetic implications described above.

Acetone, like most aliphatic ketones, has two tautomeric forms which are shown below:

$$\underset{\text{keto}}{CH_3 - \overset{\overset{\displaystyle O}{\|}}{C} - CH_3} \qquad \qquad \underset{\text{enol}}{CH_3 - \overset{\overset{\displaystyle OH}{|}}{C} = CH_2}$$

In regard to the reaction of acetone with I_2, the enol form is very important because I_2 readily reacts with it whereas it does not react with the keto form.

The presence of an acid assists in the interconversion of the keto and enol forms of acetone. It does not, of course, influence the proportions of the two forms present, but it causes an acceleration in the reactions achieving tautomerisation. It does this by making possible a new mechanism for the interconversion of the two forms, in which a proton is first added to and then removed from the acetone molecule. These steps are indicated below:

$$CH_3COCH_3 + H^+ \underset{-1}{\overset{1}{\rightleftharpoons}} CH_3\overset{\overset{+OH}{\|}}{C} CH_3$$

$$CH_3\overset{\overset{+OH}{\|}}{C} CH_3 \overset{2}{\longrightarrow} CH_3\overset{\overset{OH}{|}}{C}:CH_2 + H^+$$

(12.15)

The most usual fate of the protonated species is to lose the proton from the O atom in step -1, giving back the keto form. By step 2, a proton is lost from one of the methyl groups, which produces the enol form. This then reacts, very rapidly, with I_2 to give the reaction products:

$$CH_3\overset{\overset{OH}{|}}{C}:CH_2 + I_2 \overset{3}{\rightarrow} CH_3COCH_2I + HI$$

(12.16)

In the absence of I_2, where step 3 is not possible, the reverse of step 2 would occur at a rate equal to that of step 2, so that the concentration of the enol form would remain constant.

To consider the kinetic implications of this mechanism, it is useful to focus on the reactions of the protonated species. In the presence of I_2, this species is involved in reactions 1, -1 and 2, being formed in the first of these and removed in the other two. If the approximation is made that, at any stage in the reaction, the rate of step 1 may be equated to the sum of those of steps -1 and 2, then it is possible to obtain an expression for the concentration of the protonated species in terms of that of the keto form and the respective rate constants for the three steps in which the protonated species is involved. This is called the Principle of Stationary States, which is an assertion that the concentration of a reaction intermediate is essentially constant, at its steady-state value.

Applying this Principle to this protonated species, and using $\mathbf{A}$ for the keto form of acetone and $\mathbf{AH}^+$ for the protonated species, we then obtain:

$$\frac{d[\mathbf{AH}^+]}{dt} = k_1[\mathbf{A}][H^+] - k_{-1}[\mathbf{AH}^+] - k_2[\mathbf{AH}^+] = 0 \tag{12.17}$$

$$\therefore [\mathbf{AH}^+] = \frac{k_1[\mathbf{A}][H^+]}{k_{-1} + k_2} \tag{12.18}$$

The rate of producing the enol form in step 2 is then obtained by substitution:

$$\frac{d[\text{enol}]}{dt} = k_2[\mathbf{AH}^+]$$

$$= \frac{k_1 k_2[\mathbf{A}][H^+]}{k_{-1} + k_2} \tag{12.19}$$

In the presence of I_2, the only fate of the enol form is to react with I_2, so the rate of forming iodoacetone is also given by Equation (12.19). So on the basis of this mechanism for Reaction (12.14), the rate is proportional to the concentration of acetone and to the concentration of H^+ and is independent of the concentration of I_2, just as was found experimentally.

One consequence of a reaction being acid catalysed, in the above manner, is that if the H^+ concentration is much greater than that of acetone, then the pseudo-first order rate constant determined from the decrease in the acetone concentration will be proportional to the concentration of H^+. As a result, the logarithm of this rate constant will be a linear function of the pH, with a slope of -1.

12.4 Catalysis by an enzyme

A great many of the chemical reactions that take place within a living organism are catalysed by substances called enzymes, whose catalytic action is extremely specific. For example, one enzyme is called lactate dehydrogenase and its sole function is to assist the dehydrogenation of the lactate anion, $CH_3CH(OH)CO_2^-$, to the pyruvate ion, $CH_3COCO_2^-$.

In terms of their composition, most enzymes are polypeptides with a particular sequence of amino acids, with a molar mass usually between 10 000 and 100 000. In some cases an enzyme contains, as an integral component, an ion of a transition metal such as Fe, Co, Mn or Zn. The properties of the enzyme also depend on the specific shape and structure of the polypeptide, which arises from attachments between different peptide links by means of hydrogen bonding or by covalent disulfide bridges between the sulfur-containing cysteine residues.

For an enzyme to act on a molecule, normally called the substrate, it is necessary for intimate contact to occur between the two. The specificity of the enzyme action is usually achieved by the need for the substrate to fit closely into a cavity within the enzyme, much as a key fits into the lock for which it was designed, but not into other locks. Only then can the catalytic function of the enzyme be brought to bear on the substrate.

The reaction model normally employed to represent the mechanism of enzyme action involves two steps. In the first of these, a reversible step, the substrate, S, and the enzyme, E, form the complex, E.S. In the second, the substrate undergoes reaction within this complex to give the product, P, which separates from the enzyme:

$$\left. \begin{array}{ccc} E + S & \underset{-1}{\overset{1}{\rightleftarrows}} & E.S \\[2ex] E.S & \overset{2}{\rightarrow} & E + P \end{array} \right\} \qquad (12.20)$$

Regarding Mechanism (12.20), two points are important. The first is that in a system like this, the quantity most easily measured is the nominal concentration of the enzyme, that is, the number of moles added divided by the volume of the solution. Of this value, some proportion is present as individual molecules of enzyme and some as enzyme molecules complexed to the substrate. It is convenient to use the symbol $[E]_o$ to denote the nominal concentration, which we may equate to the sum of these two contributions:

$$[E]_o = [E] + [E.S] \tag{12.21}$$

The same point also applies to the substrate, but here it has negligible significance. In a typical experiment, the nominal concentration of the enzyme will be very much less than that of the substrate, perhaps by two orders of magnitude. Thus the concentration of the complex, E.S, will be comparable to that of the enzyme, E, but very much less than that of the substrate, S. In the light of this, the distinction between the actual and the nominal concentrations of the substrate can safely be ignored.

Applying the Principle of Stationary States to the complex in Mechanism (12.20), we have:

$$\frac{d[E.S]}{dt} = k_1[E][S] - k_{-1}[E.S] - k_2[E.S] = 0 \tag{12.22}$$

$$\therefore [E.S] = \frac{k_1[E][S]}{k_{-1} + k_2} \tag{12.23}$$

Since the actual concentration of the enzyme is not readily measured, it is appropriate to substitute for it using Equation (12.21). This gives, after a little bit of algebra,

$$[E.S] = \frac{[E]_o[S]}{\dfrac{k_{-1} + k_2}{k_1} + [S]} \tag{12.23a}$$

which can be written more neatly as:

$$[E.S] = \frac{[E]_o[S]}{K_m + [S]} = \frac{[E]_o}{1 + K_m/[S]} \tag{12.23b}$$

where $K_m = (k_{-1} + k_2)/k_1$, is known as the Michaelis constant. An expression for the rate of producing the product, P, can now be written:

$$
\begin{aligned}
\text{Rate} = v &= k_2[E.S] \\
&= \frac{k_2[E]_o[S]}{K_m + [S]} \\
&= \frac{k_2[E]_o}{1 + \dfrac{K_m}{[S]}}
\end{aligned}
\tag{12.24}
$$

This is known as the Michaelis–Menten equation and it demonstrates how the rate should vary with the concentration of the enzyme and of the substrate. Whereas it predicts that the rate is proportional to the nominal concentration of the enzyme, the dependence on the substrate concentration is rather different, leading to an increase towards a plateau value as the concentration of the substrate is increased, as shown in Figure 12.1. Where the nominal concentration of the enzyme is fixed, the greatest value the rate can approach is given by $k_2[E]_o$, which is sometimes called v_{max}. The concentration of S at which the rate attains half of this eventual maximum value is equal to K_m, the Michaelis constant.

By taking the reciprocal of each side of Equation (12.24), we obtain,

$$\frac{1}{v} = \frac{1}{k_2[E]_o}\left(\frac{K_m}{[S]}+1\right)$$

$$= \frac{1}{v_{max}}\left(\frac{K_m}{[S]}+1\right) \tag{12.25}$$

which shows that when v^{-1} is plotted against $[S]^{-1}$, we should obtain a straight line. The intercept of this line will be equal to $1/v_{max}$ and the slope to K_m/v_{max}. Where the nominal concentration of the enzyme is known, both K_m and k_2 can be evaluated from such a graph.

In studies on the effect of substrate concentration on the rate of an enzyme-catalysed reaction, it is normal to utilise the initial rates of reaction in a series of experiments in which the concentration of the substrate is varied widely, but subject always to the constraint, as explained above, that the nominal concentration of the substrate is much greater than that of the enzyme.

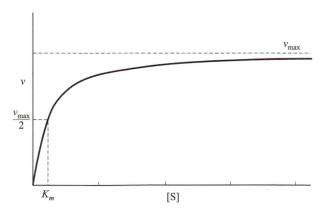

Figure 12.1 Plot of v against the concentration of the substrate, S, for an enzyme-catalysed reaction. The hypothetical maximum value of v, towards which it is rising with increasing substrate concentration, and the value of [S] at which v has attained half this value, are shown with the dotted lines.

In the above discussion, the Michaelis constant, K_m, appears as a shorthand way of writing a ratio involving the rate constants of three reaction steps. It can also be shown to have a certain physical significance. The first part of Mechanism (12.20) represents the dynamic equilibrium in regard to the formation of the enzyme–substrate complex, E.S. The appropriate equilibrium constant, if written as the constant for the *dissociation* of E.S, is given by k_{-1}/k_1. Algebraically, K_m differs from this only in that it has an additional second term in the numerator. It follows that K_m is greater than this dissociation constant, but where $k_2 \ll k_{-1}$, as is sometimes the case, the difference will be very slight. So we can say that K_m, as evaluated from the kinetic studies outlined above, will be a maximum value for the dissociation constant of the complex E.S.

Another term that is sometimes used to indicate the efficiency of an enzyme for a particular reaction is the **turnover number**. This is the number of molecules caused to react in unit time per molecule of enzyme, and optimum conditions are assumed. Mathematically, one divides v_{max} by $[E]_0$, so that the turnover number is equal to the rate constant, k_2.

12.5 Enzyme inhibition

Some molecules, such as homologues, may show sufficient structural resemblance to the substrate so that, to use the simile of the previous section, they are capable of playing the role of a key which also may fit into the lock. That is to say, this molecule also can form a complex with the enzyme. However, while it resembles the substrate, it is not the substrate: consequently, this rogue molecule will be unable to undergo reaction catalysed by the enzyme. But while it continues to play the role of a key in the lock, its presence must deny this service to the substrate and so it inhibits the catalytic action of the enzyme.

For example, malonic acid, $HO_2CCH_2CO_2H$, differs from succinic acid, $HO_2C(CH_2)_2CO_2H$, by having one less methylene group between the carboxylates.

$$
\begin{array}{ll}
\begin{array}{l}
CO_2H \\
| \\
CH_2 \\
| \\
CO_2H
\end{array}
\quad \text{malonic acid}
&
\begin{array}{l}
CO_2H \\
| \\
CH_2 \\
| \\
CH_2 \\
| \\
CO_2H
\end{array}
\quad \text{succinic acid}
\end{array}
$$

The enzyme succinate dehydrogenase acts on the succinate ion and catalyses the removal of two H atoms from it. The malonate ion shows sufficient resemblance to form a complex with this enzyme, but is unable to undergo the dehydrogenation reaction. Thus, by occupying the site on the enzyme, it acts as an inhibitor.

To set up an appropriate model for such a system, we need to add to Mechanism (12.20) the reversible step,

$$E + I \quad \rightleftharpoons \quad E.I \tag{12.26}$$

where I represents a molecule of the inhibitor and E.I its complex with the enzyme.

The situation portrayed in this revised model is usually described as the competitive inhibition of the enzyme, since the substrate, S, and the inhibitor, I, are freely competing for the available molecules of the free enzyme.

If we denote the *dissociation* constant of E.I by K_I, then we have,

$$K_I = \frac{[E][I]}{[E.I]} \tag{12.27}$$

so that the actual concentration of this complex with the inhibitor is given by:

$$[E.I] = \frac{[E][I]}{K_I} \tag{12.28}$$

Now that some molecules of the enzyme are complexed with the inhibitor, we need to replace Equation (12.21) by the relation:

$$[E]_o = [E] + [E.S] + [E.I]$$
$$= [E.S] + [E]\left\{1 + \frac{[I]}{K_I}\right\} \tag{12.29}$$

Equation (12.23) may be written as:

$$[E.S] = \frac{[E][S]}{K_m} \tag{12.30}$$

By using this equation to substitute for the concentration of the free enzyme in Equation (12.29), we obtain:

$$[E]_o = [E.S]\left\{1 + \frac{K_m}{[S]}\left(1 + \frac{[I]}{K_I}\right)\right\} \tag{12.31}$$

So the equation which, like (12.23b) in the absence of the inhibitor, expresses the actual concentration of the enzyme–substrate complex in terms of the nominal concentration of the enzyme, along with the concentrations of the substrate and the inhibitor, is:

$$[E.S] = \frac{[E]_o}{1 + \frac{K_m}{[S]}\left(1 + \frac{[I]}{K_I}\right)} \tag{12.32}$$

For the rate of the reaction under competitive inhibition, we now have:

$$v = k_2[\text{E.S}]$$

$$= \frac{k_2[\text{E}]_o}{1 + \dfrac{K_m}{[\text{S}]}\left(1 + \dfrac{[\text{I}]}{K_\text{I}}\right)}$$

$$= \frac{v_{\max}}{1 + \dfrac{K_m}{[\text{S}]}\left(1 + \dfrac{[\text{I}]}{K_\text{I}}\right)} \tag{12.33}$$

By comparing this equation with the corresponding relation, (12.24), in the absence of the inhibitor, we can see that the denominator of (12.33) is greater so that the rate of the inhibited reaction must be less. The extent of the difference obviously depends on the concentration of the inhibitor. As an illustration of the effect, if the concentration of the inhibitor is numerically equal to K_I, then at a substrate concentration equal to K_m, the rate of this inhibited reaction is now only one-third of the maximum value it can attain at very high concentrations of the substrate, not one-half as for the uninhibited reaction. As Figure 12.2 illustrates, the rate of the competitively inhibited reaction rises, with increasing concentration of the substrate, towards just the same limit as for the uninhibited reaction, namely $v = k_2[\text{E}]_o$.

By inverting Equation (12.33), we obtain,

$$\frac{1}{v} = \frac{1}{k_2[\text{E}]_o}\left\{1 + \frac{K_m}{[\text{S}]}\left(1 + \frac{[\text{I}]}{K_\text{I}}\right)\right\}$$

$$= \frac{1}{v_{\max}}\left\{1 + \frac{K_m}{[\text{S}]}\left(1 + \frac{[\text{I}]}{K_\text{I}}\right)\right\} \tag{12.34}$$

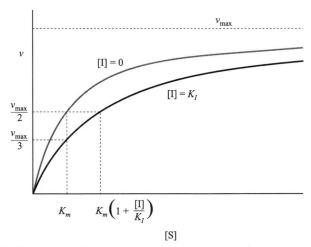

Figure 12.2   Plot of v against the substrate concentration for an enzyme-catalysed reaction in the presence of an inhibitor, I, at a concentration equal to the constant, K_I. The corresponding curve in the absence of the inhibitor (Equation 12.24) is the grey line, rising towards the same eventual limit.

which shows that for the competitively inhibited reaction, with the concentration of the inhibitor kept constant, v^{-1} should be a linear function of the reciprocal of the substrate concentration. On this plot, illustrated in Figure 12.3, the intercept is exactly the same as that of the uninhibited reaction, but the slope is greater than in the absence of inhibition, by the factor $(1 + [I]/K_I)$.

While the model consisting of reaction steps (12.20) and (12.26) can adequately describe the behaviour of some enzyme inhibitors, in some other cases the implications of Equations (12.33) and (12.34) are found not to be observed. Reaction (12.26) conceives of the formation of the enzyme–inhibitor complex as being a reversible process, in dynamic equilibrium. The most dangerous inhibitors of enzyme action tend to attach themselves so firmly to the enzyme that this process cannot be reversed, with the consequence that the enzyme molecule thereby becomes incapacitated. In this event, no matter how much substrate may be added, it will never be possible to attain the same maximum rate as in the absence of the inhibitor. This occurs because the consequence of adding a small amount of an irreversible inhibitor is effectively to remove the corresponding amount of enzyme.

An example of the irreversible attachment of an inhibitor to an enzyme is provided by the attack of diisopropyl fluorophosphate on acetylcholinesterase, in which this foreign substance bonds covalently to the enzyme close to the active cite, thus rendering the enzyme ineffective. Acetylcholine plays a crucial role as a chemical transmitter between nerve endings and muscles, so the incapacitation of this enzyme causes serious dysfunction to an animal body. In effect, diisopropyl fluorophosphate is a lethal nerve gas.

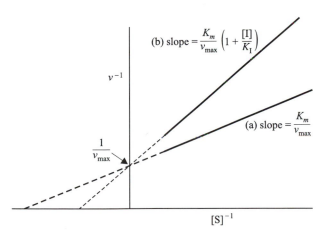

Figure 12.3 Plots of v^{-1} against $[S]^{-1}$ for enzyme reactions (a) without an inhibitor present and (b) with a constant concentration of the inhibitor, I, illustrating that Equations (12.25) and (12.34) predict the same intercept (v_{max}^{-1}) but different slopes for these two plots.

12.6 The influence of temperature on enzyme action

The effect of temperature variation on an enzyme-catalysed reaction is a little more complex than that on most chemical reactions. Over a range of temperatures, a fairly normal increase is usually found. But at yet higher temperatures, this increase is not sustained and the rate is found typically to go through a maximum, above which it declines fairly sharply. The usual explanation for this maximum at what is called the optimum temperature is that enzymes, being peptide molecules, are liable to undergo denaturation at temperatures some way above their normal operational temperature, and are thus rendered inactive.

If experimental measurements are confined to temperatures well below the optimum temperature, in a range where the rate of the denaturation process is negligible, then it is possible to deduce information about the enzyme-catalysed reaction. For a simple reaction of this type, occurring by the Michaelis–Menten mechanism, the rate expression is given by Equation (12.24). At high substrate concentrations, this tends towards the limiting value given by the equation,

$$v_{max} = k_2[E]_o \qquad (12.35)$$

where k_2 denotes the rate constant of the step in which the enzyme–substrate complex reacts to yield the product(s). It follows that measurements of the temperature dependence of v_{max}, carried out under these conditions, should yield the activation energy of this particular step, which we shall call E_2. This will be obtained as $-R$ times the slope of the Arrhenius plot of $\ln v_{max}$ against T^{-1}.

Alternatively, it is possible to conduct the rate experiments under conditions of very low substrate concentration, where $[S] \ll K_m$. We then have:

$$
\begin{aligned}
v &= \frac{k_2[E]_o[S]}{K_m} \\
&= \frac{k_1 k_2[E]_o[S]}{(k_{-1} + k_2)}
\end{aligned} \qquad (12.36)
$$

Desirably, the relative values of k_{-1} and k_2 should be known. Where one is known to be much greater than the other, then it is permissible to simplify further: but if k_{-1} and k_2 are comparable or their relative values are unknown, then the effect of temperature on v under these conditions is impossible to interpret.

If $k_{-1} \gg k_2$ then we have,

$$v = \frac{k_1 k_2}{k_{-1}}[E]_o[S] \qquad (12.37)$$

and any activation energy deduced from the variation of v with temperature has to be attributed to the function, $k_1 k_2 / k_{-1}$. If we assume that each of these individual rate constants obeys the Arrhenius equation (see section 11.6), with a correspondingly numbered activation energy, then we have:

$$-R \cdot \frac{d\ln(k_1 k_2/k_{-1})}{d(1/T)} = E_1 + E_2 - E_{-1} \tag{12.38}$$

In this case, the variation of the rate with the temperature is a function of the activation energies of all three reaction steps, 1, 2 and -1.

If k_{-1} and k_2 are comparable, but with quite different activation energies, then the plot of $\ln v$ against T^{-1} is liable to be appreciably curved, with a mean slope corresponding to no function in particular.

12.7 Reactions of free radicals *in vivo*

In the normal metabolic processes, stable molecules (such as sugars, fats or peptides) undergo a series of degradative reactions, each catalysed by the appropriate enzyme, to produce stable molecules as products. However, it is possible for unstable species known as free radicals to be generated within an animal system. In these, the normal valency of at least one constituent atom is not satisfied, so that the entity will possess an unpaired electron and is usually very reactive.

It may be pointed out, parenthetically, that the heterolytic dissociation of a molecule does not have this effect, because it does not lead to the valency of any constituent atom being infringed. If the acid, RCO_2H, ionises to give RCO_2^- and H^+, there is now a formal charge of -1 on one of the O atoms, making it isoelectronic with F, which is univalent. So, in the formula,

$$R-C\overset{\displaystyle O^-}{\underset{\displaystyle O}{\big\backslash}}$$

every valency is satisfied. Also, this species has an even number of electrons, so that no unpaired electron has been created. (The sharing of the electronic charge between the two O atoms in the carboxylate group has been referred to in Chapter 9. This elaborates the picture, but does not alter it.)

As an example of a free radical, consider the hydroxyl radical, $\dot{O}H$, in which the O atom is bonded only to one H atom, so that there is a univalent oxygen atom. The dot above the letter O denotes this free radical centre, with its unpaired electron. The reactivity of the hydroxyl radical is attested by its tendency to react with any organic compound containing an alkyl group. In this process, the hydroxyl radical acquires an H atom to become a water molecule, while the organic molecule, by losing the H atom, has become a free radical:

$$\dot{O}H + CH_3CH_2OH \rightarrow H_2O + CH_3\dot{C}HOH \qquad (12.39)$$

In the light of the large numbers of alkyl H atoms in sugars, fats and peptides, this means that all of these molecules are the potential target of destructive attack by free radical species. Enzymes are peptides and are thus vulnerable. Deoxyribonucleic acid (DNA) contains sugar residues, and must be liable to attack if $\dot{O}H$ radicals should be generated in its vicinity.

In addition, the hydroxyl radical is a strong oxidising agent. It can, for example, oxidise Fe^{2+} to Fe^{3+}, being itself reduced to the hydroxide ion in the process:

$$\dot{O}H + Fe^{2+} \rightarrow OH^- + Fe^{3+} \qquad (12.40)$$

This means it has the ability to upset the oxidation state of the transition metal ion in a number of important enzymes.

Among free radicals *in vivo*, a most important species is the hydroperoxy radical, $HO_2^{\cdot}$. This behaves as a weak acid,

$$HO_2^{\cdot} \rightleftharpoons H^+ + O_2^{-} \qquad (12.41)$$

with a pK_a value of 4.9, so that at the pH of most biological media, it is predominantly present as the anion O_2^{-}. This is a species with an odd number of electrons, so that it is both an ion and a free radical, hence the term, the superoxide radical anion.

Superoxide may be generated by any species that is able to reduce the dioxygen molecule:

$$O_2 + X^- \rightarrow O_2^{-} + X^{\cdot} \qquad (12.42)$$

For this process, a possible reducing agent is an Fe^{2+} ion, complexed with an appropriate chelating group. Additionally, O_2 may be reduced to O_2^{-} in the course of the photo-oxidation of the amino-acid tryptophan, by near-UV light. The significance of tryptophan in this connection is that it is one of the few amino acids to absorb light in this wavelength range and the only one containing the indole moiety.

In aqueous solution, superoxide undergoes the disproportionation reaction which we may write stoichiometrically as:

$$HO_2^{\cdot} + HO_2^{\cdot} = O_2 + H_2O_2 \qquad (12.43)$$

Since this process occurs most rapidly at pH 5, the relevant reaction step may be deduced to be,

$$HO_2^{\cdot} + O_2^{-} \rightarrow HO_2^{-} + O_2 \qquad (12.44)$$

followed by the protonation step:

$$HO_2^{-} + H^+ \rightarrow H_2O_2 \qquad (12.45)$$

Although Reaction (12.44) is known to have a rate constant of 8×10^7 dm^3 mol^{-1} s^{-1}, which is a fairly high figure for a bimolecular reaction, it cannot serve as an effective path for the destruction of superoxide *in vivo*. One reason is that, being a second order reaction, the rate of Reaction (12.44) decreases by a factor of four as the concentration of superoxide is halved. Another is that, at the relevant biological pH values, superoxide exists almost exclusively as the O_2^- ion, and the rate constant for the charge transfer process,

$$O_2^- + O_2^- \rightarrow O_2 + O_2^{2-} \tag{12.46}$$

is virtually negligible.

In 1968, an enzyme was discovered whose only known function is to catalyse the dismutation of superoxide, so it is called superoxide dismutase (SOD). This enzyme contains ions of both copper and zinc, and its catalysis of Reaction (12.43) is essentially independent of pH, at least over the range pH 5.3 to 9.5. At low concentrations of superoxide, the rate of the dismutation is proportional to the concentrations of superoxide and of SOD, so that the efficiency of the catalysed reaction can be encapsulated in a second order rate constant, whose value is around 2×10^9 dm^3 mol^{-1} s^{-1}. This high value means that the reaction of the superoxide radical ion with the SOD enzyme is within a factor of five of being as fast as the transport of the two species through an aqueous medium would allow.

The disproportionation reaction of superoxide involves the oxidation of one radical anion, accompanied by the reduction of another. It seems probable that the copper ion in SOD facilitates this process, by alternately undergoing reduction and oxidation between Cu^{2+} and Cu^+, i.e.,

$$\left. \begin{array}{rcl} E{-}Cu^{2+} + O_2^- & \rightarrow & E{-}Cu^+ + O_2 \\ E{-}Cu^+ + O_2^- + 2H^+ & \rightarrow & E{-}Cu^{2+} + H_2O_2 \end{array} \right\} \tag{12.47}$$

where **E** denotes the remainder of the enzyme. The protons needed in the second step may be supplied by histidine residues in the peptide part of the enzyme, wrapped around the copper ion, thus explaining the independence of the rate on the pH.

12.8 The role of antioxidants

Hydrogen peroxide is toxic within the cell and is removed by the action of various enzymes. However, where it is present, superoxide may react with it to generate the hydroxyl radical:

$$O_2^- + H_2O_2 \rightarrow O_2 + OH^- + \dot{O}H \tag{12.48}$$

This reaction may be catalysed by an ion of a transition metal, such as iron, even it is complexed. The mechanistic steps would then be:

$$O_2^- + Fe^{3+} \quad \rightarrow \quad O_2 + Fe^{2+}$$
$$Fe^{2+} + H_2O_2 \quad \rightarrow \quad Fe^{3+} + \dot{O}H + OH^-$$

$$\left. \right\} \qquad (12.49)$$

The hydroxyl radical is capable of abstracting an H atom from an alkyl group, and in lipids this may, in the presence of oxygen, lead to chain per-oxidation. The sequence starts with the creation of a free radical centre on the alkyl group:

$$\dot{O}H + \overset{\backslash}{\underset{/}{C}}H_2 \quad \rightarrow \quad H_2O + \overset{\backslash \cdot}{\underset{/}{C}}H \qquad (12.50)$$

Where O_2 is present, this radical may react with it to yield an organic peroxide,

$$\overset{\backslash \cdot}{\underset{/}{C}}H + O_2 \quad \rightarrow \quad \overset{\backslash}{\underset{/}{C}}H\dot{O}_2 \qquad (12.51)$$

which is capable of abstracting a hydrogen atom from a neighbouring alkyl group:

$$\overset{\backslash}{\underset{/}{C}}H\dot{O}_2 + \overset{\backslash}{\underset{/}{C}}H_2 \quad \rightarrow \quad \overset{\backslash}{\underset{/}{C}}HO_2H + \overset{\backslash \cdot}{\underset{/}{C}}H \qquad (12.52)$$

Taken together, Reactions (12.51) and (12.52) constitute chain peroxidation, where these two reactions are the propagation steps.

The term **antioxidant** is widely used to denote a substance which inhibits an oxidation process such as that detailed above. Vitamin E, α-tocopherol, which has been shown to be essential for the good health of a range of animals, is known to be an antioxidant. The hyrdroxyl radical reacts readily with it, as does the superoxide radical anion. However, *in vivo* there is present so much more lipid than α-tocopherol that although the presence of this antioxidant provides a reaction mode for the $\dot{O}H$ radical which is in competition with Reaction (12.50), it can only have a negligible effect on the extent of that reaction.

But α-tocopherol (TH) can also react with the peroxy radical formed in Reaction (12.51):

$$\overset{\backslash}{\underset{/}{C}}H\dot{O}_2 + TH \quad \rightarrow \quad \overset{\backslash}{\underset{/}{C}}HO_2H + T^{\bullet} \qquad (12.53)$$

The α-tocopherol radical, $T^{\bullet}$, formed in this step, is not sufficiently reactive to abstract an H atom from a lipid, so that Reaction (12.53) represents chain termination. The radical, $T^{\bullet}$, may be reduced back to TH by ascorbic acid, vitamin C, so that the participation of α-tocopherol in Reaction (12.53) need not be sacrificial. In this way, vitamin E is all the more effective as an inhibitor of the oxidation of lipids.

Vitamin E

Suggested reading

LOGAN, S. R., 1996, *Fundamentals of Chemical Kinetics*, London: Longman. (Chapters 3 and 9).

PISZKIEWICZ, D., 1977, *Kinetics of Chemical and Enzyme-catalyzed Reactions*, New York: Oxford University Press.

HALLIWELL, B. and GUTTERIDGE, J. M. C., 1989 *Free Radicals in Biology and Medicine*, 2nd Edn, Oxford: Clarendon.

Problems

12.1 An alkyl halide, RX, undergoes hydrolysis in an aqueous medium by both an S_N1 and an S_N2 mechanism, with respective rate constants of k_1 and k_2. In each experiment, the initial concentration of RX is kept low, so that, regardless of the pH, the RX concentration decreases in an exponential manner. Express the relevant pseudo-first order rate constant, k', for the removal of RX in terms of k_1, k_2 and whatever additional parameters are necessary.

12.2 For the oxidation of Tl^+ by Ce^{4+},

$$2Ce^{4+} + Tl^+ = 2Ce^{3+} + Tl^{3+}$$

catalysed by Ag^+ ion, the following mechanism has been proposed:

$$Ce^{4+} + Ag^+ \underset{-1}{\overset{1}{\rightleftarrows}} Ce^{3+} + Ag^{2+}$$

$$Ag^{2+} + Tl^+ \overset{2}{\rightarrow} Ag^+ + Tl^{2+}$$

$$Tl^{2+} + Ce^{4+} \overset{3}{\rightarrow} Tl^{3+} + Ce^{3+}$$

By applying the Principle of Stationary States to Ag^{2+}, derive an expression for the rate of this catalysed reaction in terms of the concentrations of Ce^{4+}, Tl^+, Ce^{3+} and Ag^+ and the rate constants k_1, k_{-1} and k_2.

12.3 The dehydration of 3-hydroxybutanoic acid (B) to form a lactone was studied in 0.2 mol dm^{-3} aqueous HCl at 298 K. The concentration of B was found to vary with time as follows:

t/min	0	21	36	50	65	80	100	∞
[B]/arb. units	100	86.8	78.9	72.5	66.5	61.2	55.3	27.1

Interpreting this dehydration as a reversible reaction, with the forward and the reverse processes both first order, determine the rate constant for each.

12.4 The hydrolysis of methyl 3-phenylpropanoate (an ester) is catalysed by the enzyme chymotrypsin. At 298 K, a pH of 7.6 and a constant concentration of the enzyme, the following data were obtained for the initial rates of hydrolysis:

[Ester]/(10^{-3} mol dm^{-3})	30.8	14.6	8.57	4.60	2.24	1.28
Initial rate/(10^{-8} mol dm^{-3} s^{-1})	20.0	17.5	15.0	11.5	7.5	5.0

Evaluate the limiting rate of reaction under these conditions and the Michaelis constant, K_m.

APPENDIX ONE

Indices and logarithms

Logarithms are a mathematical device which enables certain relationships to be expressed in a succinct way. They are used in mathematical equations such as those that interrelate certain quantities employed in physical chemistry. To understand how logarithms behave, it is desirable to have a thorough knowledge of indices, so that will be our starting point.

In the expression x^n, the number n, written as a superscript, is called an index. If n is a positive integer, then by x^n we mean x times x times x etc., until we have n factors. If we now multiply x^n by x^m, where m is another positive integer, we have as follows:

$$x^n \cdot x^m = \underbrace{x.x.x. \ldots x}_{n \text{ factors}} \cdot \underbrace{x.x.x \ldots x}_{m \text{ factors}}$$

$$= \underbrace{x.x.x. \ldots\ldots x}_{(n+m) \text{ factors}} \tag{A1.1}$$

$$= x^{n+m}$$

So, when we multiply, we add the indices.

In a similar way, we may treat division:

$$\frac{x^n}{x^m} = \frac{\overbrace{x.x.x.....x}^{n \text{ factors}}}{\underbrace{x.x.x.....x}_{m \text{ factors}}}$$

$$= \underbrace{x.x.x.....x}_{(n-m) \text{ factors}} \tag{A1.2}$$

$$= x^{n-m}$$

(For convenience, we have assumed that n is greater than m). This shows that when we divide, we subtract the indices.

In deriving Equations (A1.1) and (A1.2), an important stipulation was that n and m were positive integers. One would wish to dispense with this restriction on the use of indices, but it must be desirable that all indices conform to the same rules. Let us now contemplate zero, negative and fractional indices on the strict basis that they satisfy equations (A1.1) and (A1.2).

Starting with zero, if we multiply x^0 by x^n we have, by adding the indices:

$$x^0 \cdot x^n = x^n \tag{A1.3}$$

We may now divide across by x^n and obtain:

$$x^0 = 1 \tag{A1.4}$$

So, regardless of x, x^0 has the value of 1.

For the negative index, let us multiply x^{-n} by x^n.

$$x^{-n} \cdot x^n = x^0 = 1 \tag{A1.5}$$

Dividing across by x^n, this gives:

$$x^{-n} = \frac{1}{x^n} \tag{A1.6}$$

So, regardless of the value of x, x^{-n} is the reciprocal of x^n.

Turning to fractional indices, such as $x^{1/2}$, if we multiply this by itself, using Equation (A1.1), we have:

$$x^{1/2} \cdot x^{1/2} = x^1 = x \tag{A1.7}$$

This means that $x^{\frac{1}{2}}$ is equal to the square root of x, that is:

$$x^{1/2} = \sqrt{x} \tag{A1.8}$$

It is a simple matter to extend this to other simple fractions, like $\frac{1}{3}$, or $\frac{1}{4}$, or $\frac{1}{5}$ or $\frac{2}{5}$.

We are now in a position to introduce the logarithm. This may be done as a simple definition:

If x is equal to a to the power of p, then the logarithm of x to the base a is equal to p.

In equation form, this may be written as:

If $x = a^p$, then $\log_a x = p$

It is of interest to look at the logarithm of a product. Suppose $x = a^p$ and $y = a^q$. We then have:

$$xy = a^p \cdot a^q$$
$$= a^{p+q} \tag{A1.9}$$
$$\therefore \log_a(xy) = p + q$$
$$= \log_a x + \log_a y \tag{A1.10}$$

So the logarithm of the product is the sum of the logarithms. Likewise, the logarithm of a quotient is the difference in the logarithms:

$$\frac{x}{y} = \frac{a^p}{a^q} = a^{p-q} \tag{A1.11}$$

$$\therefore \log_a \left(\frac{x}{y}\right) = p - q$$

$$= \log_a x - \log_a y \tag{A1.12}$$

In the light of Equations (A1.10) and (A1.12), logarithms can have an important role in calculations involving multiplication and division. They were extensively used for this purpose before the era of the electronic calculator.

Numbers used as a base for logarithms include 10, e and 2. We will now look at some numerical examples using 10 as the base. For powers of ten, positive or negative, the logarithm is easily evaluated, and a few are shown below.

Number			Logarithm to the base 10
1000	=	10^4	4
1000	=	10^3	3
100	=	10^2	2
10	=	10^1	1
1	=	10^0	0
0.1	=	10^{-1}	−1
0.01	=	10^{-2}	−2
0.001	=	10^{-3}	−3

The data in this table are plotted in Figure A1.1, from which it may be seen that the plot of $\log_{10} x$ against x is a characteristic curve, whose slope is everywhere positive but continually decreasing as x increases. When $x > 1$, the logarithm is positive, whereas where $x < 1$, the logarithm is negative. However, it is the logarithm and not the number that is negative: we have no logarithms for negative values of x.

The square root of 10 is 3.162, and since this number is $10^{0.5}$, we have:

$$\log_{10} 3.162 = 0.50 \tag{A1.13}$$

Since

$$3162 = 3.162 \times 10^3 \tag{A1.14}$$
$$\log_{10} 3162 = 0.50 + 3.0$$
$$= 3.50 \tag{A1.15}$$

Alternatively, 0.03162 is 3.162×10^{-2}, so its logarithm to the base 10 is 0.50 −2.0 = −1.50. In short, if we are given any number between 1 and 10 whose logarithm we know, we can readily generate the logarithm of that number multiplied by or divided by any power of 10. Herein lies the case for using logarithms to the base 10.

The number e may be expressed as the sum of the infinite series,

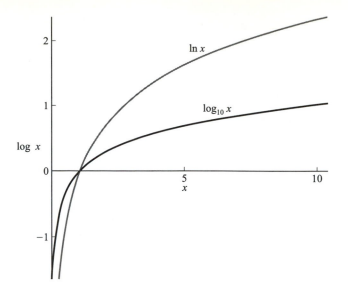

Figure A1.1 Plots of $\log_{10}x$ and of $\log_e x(= \ln x)$ against x.

$$e = 1 + \frac{1}{1!} + \frac{1}{2!} + \frac{1}{3!} + \frac{1}{4!} + \cdots = 2.7182818\cdots \tag{A1.16}$$

which is rapidly convergent because, after a certain point, each term is less than its predecessor by an ever increasing factor.

Regarding the use of this number as a base for logarithms, two points are particularly relevant:

1 This relates to calculus, and may more readily be comprehended in the context of the next section. Differentiation of a term in x^n yields a term in x^{n-1}. One might then wonder what function of x can yield, on differentiation, a term in x^{-1}. The answer is that this function is $\log_e x$. Consequently, the integral of $x^{-1}dx$ yields a term in $\log_e x$.

2 Logarithms to the base e may be written as an infinite series:

$$\log_e(1 + x) = x - \frac{x^2}{2} + \frac{x^3}{3} - \frac{x^4}{4} + \cdots \tag{A1.17}$$

This means that $\log_e(1 + x) \to x$ as $x \to 0$, so that $\log_e(1 + x) \approx x$ is an acceptable approximation if x is very small.

For these and other reasons, it is mathematically convenient to take logarithms to the base e, which may be regarded as the natural base for that purpose. Among chemists, the symbol used is 'ln', read as 'natural logarithm'.

$$\ln x \equiv \log_e x \tag{A1.18}$$

If the logarithmic base is changed, then a simple conversion factor is applicable.

$$\log_a x = \log_p x \cdot \log_a p \tag{A1.19}$$

Regarding the bases 10 and e, this relation becomes,

$$\ln x = 2.303 \log_{10} x \tag{A1.20}$$

where 2.303 is the numerical value of $\ln 10$. The interrelation of logs to the base e and to the base 10 is shown graphically in Figure A1.1. At $x = 1$, the two coincide since they are both zero. Where x exceeds one, $\ln x$ is greater than $\log_{10} x$, but where x is less than one, both are negative and $\ln x$ is more negative than $\log_{10} x$.

APPENDIX TWO

Introduction to calculus

Calculus is a branch of mathematics which is ideal for handling the concept of the variation of one parameter as a function of changes in another. It was first developed by Leibnitz, but when Newton was working towards his Second Law of Motion, he saw the need for this branch of mathematics and, being unaware of the earlier work, he developed his own version of calculus.

It is convenient to start with the equation,

$$y = mx + c \tag{A2.1}$$

which means that when $x = 0$, $y = c$, and when $x = -m/c$, $y = 0$. This is an equation for a straight line through these two points, and we may say that when an equation of this form is applicable, y is a linear function of x.

For the special case where c was equal to zero, then the line would pass through the origin and we could say that y is proportional to x (though we would do so only if m were positive).

A straight line is characterised by its slope and if we take two points on such a line and label them 1 and 2 then the slope is given by:

$$\text{slope} = \frac{y_2 - y_1}{x_2 - x_1} \tag{A2.2}$$

For the line corresponding to Equation (A2.1), the slope is equal to m and the intercept on the y axis is equal to c.

Let us consider the function,

$$y = x^2 + 6 \tag{A2.3}$$

which is sketched in Figure A2.1. We will choose a point on this curve and call its co-ordinates x and y. Let us now take a value of x a little greater than this, and call it $(x + \delta x)$, where the Greek letter δ (delta) indicates a small change, so that δx represents an incremental increase in x. As a consequence of this small change in the value of x, that of y has changed to $(y + \delta y)$, so that the point, $(x + \delta x, y + \delta y)$ lies on the curve, not far from the point, (x, y).

We wish to look at the ratio of the incremental change in y to that in x which produces it, that is, in $\delta y/\delta x$. Since this ratio depends on how large or small we make δx, we are more specifically interested in the limiting ratio, as δx

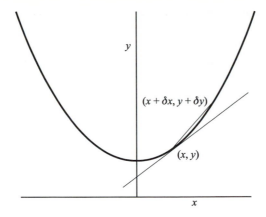

Figure A2.1 A plot of $y = x^2 + 6$ against x, showing the chord through the points (x, y) and $(x + \delta x, y + \delta x)$ on this curve, and the tangent to the curve at the point (x, y).

is made vanishingly small. This will depend on the value chosen for x, but on nothing else. So we have, as a definition,

$$\frac{dy}{dx} = \frac{\lim}{\delta x \to 0}\left(\frac{\delta y}{\delta x}\right) \tag{A2.4}$$

where the left-hand side of the equation is read, 'dy by dx'.

To evaluate the right-hand side of Expression (A2.4), we can use the equation for the curve.

$$
\begin{aligned}
\delta y &= (y + \delta y) - (y) \\
&= [(x + \delta x)^2 + 6] - (x^2 + 6) \\
&= x^2 + 2x.\delta x + (\delta x)^2 + 6 - x^2 - 6 \\
&= 2x.\delta x + (\delta x)^2
\end{aligned} \tag{A2.5}
$$

This then gives, for the ratio of δy to δx:

$$\frac{\delta y}{\delta x} = 2x + \delta x \tag{A2.6}$$

As $\delta x \to 0$, the second of these terms vanishes, so we have:

$$\frac{dy}{dx} = \frac{\lim}{\delta x \to 0}(2x + \delta x) \tag{A2.6a}$$

Referring to Figure A2.1, the line drawn through the points (x, y) and $(x + \delta x, y + \delta y)$ has a slope equal to $\delta y/\delta x$. As δx is made smaller, the line ceases to be a chord and eventually becomes a tangent to the curve at (x, y), so that dy/dx represents the slope of the tangent.

One application of this process, which is called differentiation, is in locating the value of x at which the slope of the tangent to the curve (which is also the slope of the curve at that point) has a particular value. Most often, the value in question is zero, since this would indicate where the curve turns around, that is, where y has a maximum or a minimum value. In relation to Figure A2.1, equation (A2.6) shows that dy/dx will be zero where $x = 0$. Referring to the curve, we can see that y has a minimum value at $x = 0$.

For functions of the form,

$$y = ax^n \tag{A2.7}$$

it is a simple matter to develop the counterparts of Equations (A2.5) and (A2.6), where n is a positive integer. The result is,

$$\frac{dy}{dx} = anx^{n-1} \tag{A2.8}$$

and this may be shown to be valid for any value of n. Of course, if $n = 0$, Equation (A2.7) is simply, $y = a$, and Equation (A2.8) shows that on differentiating this we obtain zero.

Also, on differentiating the sum of two terms, we obtain the sum of the differentials of these two terms. That is, if

$$y = u + v \tag{A2.9}$$

where u and v are both functions of x, then

$$\frac{dy}{dx} = \frac{du}{dx} + \frac{dv}{dx} \tag{A2.10}$$

To illustrate the uses of differentiation, let us consider the function,

$$y = x^3 - 3x^2 - 9x + 6 \tag{A2.11}$$

and assess whether it exhibits any maxima or minima. We might do this by evaluating y for various integral values of x and plotting the curve, but an answer can be obtained much more quickly by employing calculus. From Equation (A2.11) we obtain:

$$\frac{dy}{dx} = 3x^2 - 6x - 9 \tag{A2.12}$$

For a maximum or a minimum, $dy/dx = 0$, so we put the right-hand side equal to zero. Factorising it, we have:

$$\begin{aligned} 3x^2 - 6x - 9 &= 3(x^2 - 2x - 3) \\ &= 3(x + 1)(x - 3) \end{aligned} \tag{A2.13}$$

So, we will have a turning point when one of the factors in equation (A2.13) is zero. The solutions are, $x = -1$ or $x = 3$. The curve of Equation (A2.11) is shown in Figure A2.2, and demonstrates that this function has a maximum at $x = -1$ and a minimum at $x = 3$.

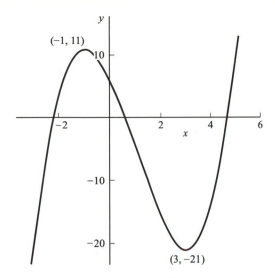

Figure A2.2 A plot of the function, $y = x^3 - 3x^2 - 9x + 6$ against x, showing that y has a minimum at $x = 3$ and a maximum at $x = -1$.

Sometimes, there is a need for the mathematical operation which is the reverse of differentiation. This is called integration. Suppose we are given an equation for the velocity of a body as a function of time. Velocity is the rate of change of position, and if position is denoted by the variable y and the time by t, then the velocity is equal to dy/dt. If we want the position at some particular time, we need the equation for y in terms of t.

It is relevant that velocity multiplied by time gives the distance travelled. This means that the distance travelled within a particular interval of time is given by the area under the curve of velocity against time, between the two pertinent values of t. To express this relation more generally, the integral of a function, obtained by reversing the differentiation process, is a measure of the area under the curve of that function plotted against its natural variable.

The result of the integration process can be deduced from the rules of differentiation, given above. If we have

$$\frac{dy}{dx} = ax^n \qquad (A2.14)$$

then it follows that y can be expressed as,

$$y = \frac{ax^{n+1}}{(n+1)} + c \qquad (A2.15)$$

where c is an unknown constant. Another way of expressing this is to say:

$$y = \int ax^n \, dx \qquad (A2.15a)$$

On the right-hand side, we have an integral sign, and then the function which we wish to integrate. The dx at the end means that we are integrating with respect to x, that is, that x is the relevant variable.

Sometimes c may be evaluated from other available information, such as a known value of y at a particular value of x. But if we are dealing with the area under the curve between specified values of x, then the desired area is the difference between the values of the integral at the upper and the lower limits, so that c cancels out. This particular evaluation is referred to as the definite integral and is denoted by having the values of the upper and lower limits of x written after the integral sign or, after having carried out the integration, writing them after the square brackets enclosing the resulting function.

Referring back to Equation (A2.3), suppose we require the area under this curve between $x = 0$ and $x = 3$. We then have:

$$\text{Area, } A = \int_0^3 (x^2 + 6)dx$$

$$= \left[\frac{x^3}{3} + 6x \right]_0^3 \tag{A2.16}$$

$$= (9 + 18) - (0)$$

$$= 27$$

Two further details of differentiation may be added, both of which have relevance to integration procedures. If y is the product of u and v, where these are both functions of x, then:

$$\frac{dy}{dx} = u\frac{dv}{dx} + v\frac{du}{dx} \tag{A2.17}$$

Alternatively, it may be convenient to consider y as a function of z, where z is some function of x. We can then differentiate using the relation:

$$\frac{dy}{dx} = \frac{dy}{dz} \cdot \frac{dz}{dx} \tag{A2.18}$$

For example, suppose we have the equation,

$$y = e^{3x^2} \tag{A2.19}$$

which we will think of as $y = e^z$, where $z = 3x^2$. The differential of e^x is e^x, so we have:

$$\frac{dy}{dx} = \frac{dy}{dz} \cdot \frac{dz}{dx}$$

$$= e^z \cdot 6x \tag{A2.20}$$

$$= 6xe^{3x^2}$$

APPENDIX THREE

Dimensions and units

Here we review some of the important units used in the physical sciences, both primary units and others derived from them. Mostly, these units belong to the internationally agreed system called SI (Système International), which has the merit that if every parameter in an equation is entered in its SI unit, then the answer will inevitably be in an SI unit. However, there are also some widely used units which do not belong to SI.

The unit of length is the metre (abbreviation, m). As for any fundamental unit, one uses multiples of this, usually in terms of the factors 10^3, 10^6, 10^{-3}, 10^{-6}, etc., by employing the relevant prefix.

Factor	Prefix	Abbreviation
10^9	giga	G
10^6	mega	M
10^3	kilo	k
10^{-3}	milli	m
10^{-6}	micro	μ
10^{-9}	nano	n
10^{-12}	pico	p

However, one non-SI unit is sometimes used to denote interatomic separations, namely the angstrom, denoted by Å, where $1\text{Å} = 10^{-10}$ m = 0.1 nm.

The unit of area is the metre squared (m^2) and that of volume is the metre cubed (m^3). In laboratory terms, the first is large and the second is huge. Whereas 1 m^3 of concrete is readily handled on a building site, very much smaller volumes of a solution suffice in the laboratory. One decimetre cubed (dm^3) is 10^{-3} m^3 and is also called the litre (l). One centimetre cubed (cm^3) is 10^{-6} m^3 and is also called the millilitre (ml). One millimetre cubed (mm^3) is 10^{-9} m^3 and is called a microlitre (μl).

For many purposes, the concentrations of solutions are reckoned in terms of the number of moles of solute in a certain volume of solution. The SI unit is one mole per m^3. In the light of the huge size of one m^3, this is quite a dilute solution. The term 'molar' pre-dates SI and refers to a concentration of one mole per dm^3 (or one mole per litre). A number of equivalent ways are used in books and journals to represent this:

233

1 mol dm^{-3} = 1 mol l^{-1} = 1 M

The recognised abbreviation of 'mole' is 'mol', which is an amount of substance. In some journals 'M' is used as the abbreviation of 'molar', the unit of concentration. In this book, concentrations are written as mol dm^{-3}, which is the style specified in the journals of the Royal Society of Chemistry.

The amount of substance dissolved within a solution can be obtained as the product of the volume and the concentration. In multiplying, one must use self-consistent units. For example, in 10.6 cm^3 of 0.1 mol dm^{-3} HCl, one has

$$(10.6 \times 10^{-3} \text{ dm}^3) \times (0.1 \text{ mol dm}^{-3}) = 1.06 \times 10^{-3} \text{ mole of HCl.}$$

One application of this relation arises in regard to the stoichiometric balance at the end-point of a titration. Suppose we are titrating an acid, such as HCl, using a base, such as NaOH. At the end-point, the respective numbers of moles of the acid and the base must be equal, so we have,

$$M_A V_A = M_B V_B \tag{A3.1}$$

where the parameters M denote the solution molarities and V the volumes of the solution and the subscripts A and B refer to the acid and the base.

The unit of mass is the kilogram (kg).

The unit of time is the second (s). Thus velocity, which is the rate of change of position, has the unit, m s^{-1}, and acceleration, which is the rate of change of velocity, has the units, m s^{-2}.

By Newton's Second Law, the acceleration produced on a body by the action of a force is proportional to the magnitude of the force and inversely proportional to the mass of the body. On that basis, we can measure the force by the product of the mass of the body and the acceleration which the force produces. Thus the unit of force, called the newton (N) is 1 kg m s^{-2}. Unit pressure is the consequence of unit force over unit area, where the latter is one square metre (1 m^2). The unit of pressure, called the pascal (Pa), is thus 1 N m^{-2}, which is the same as 1 kg m^{-1} s^{-2}.

A force acting through a distance does work, of an amount equal to the product of these two parameters. The unit of work, 1 newton-metre, is called the joule (J) and is equal to 1 kg m^2 s^{-2}. An alternative formulation is that unit work may be done if a piston in a cylinder has unit pressure exerted on it so that it moves through unit volume. (This may be seen as the consequence of dividing the force by the cross-sectional area of the cylinder to obtain the pressure and multiplying the distance the piston moves by the same area to obtain the volume.)

For electrical units, the starting point is that of electric current, called the ampere (A). This is defined on the basis of the phenomenon of electromagnetic induction, where the effect is proportional to the electric current flowing in a conductor. If a current is to flow, there must be a difference in the electrical potential; also, for current to flow, electrical work must be done. The unit of potential difference, the volt (V), is defined so that when a current of 1 A flows for 1 second between two points whose potential difference is 1 V, the amount of work done is 1 J.

The unit of the quantity of electrical charge, the coulomb (C), is that conveyed by a current of 1 A in 1 second. Electrical charges interact in a manner which depends on whether they are like (both positive or both negative) or unlike (one positive and one negative). The former mutually repel, the latter mutually attract. If we assume point charges of magnitude q_1, and q_2, at a separation r, then in a vacuum, the force of their interaction is given by:

$$F = \frac{q_1 q_2}{4\pi\varepsilon_0 r^2} \tag{A3.2}$$

where ε_0 represents the permittivity of a vacuum. This is one of the various inverse square dependences found in nature: if r were to be doubled, the force of attraction or repulsion would be decreased by a factor of four.

Where the point charges are not in a vacuum, Equation (A3.2) needs to be slightly modified, to:

$$F = \frac{q_1 q_2}{4\pi\varepsilon_0 K_r r^2} \tag{A3.3}$$

The value of the dimensionless constant K_r depends on the identity of the medium around the two charges, and it is called the relative permittivity of this medium. As examples, for *n*-hexane, $K_r = 1.9$ whereas for ethanol, $K_r = 24.3$. Clearly, ethanol diminishes the forces of interaction between the charges to a much greater extent than does *n*-hexane.

The smallest 'packet' of electrical charge in nature is that of the electron and it has the value, 1.602×10^{-19} C. The extreme smallness of this number reflects the fact that the SI units we have met are macroscopic, totally unsuited to the scale of individual atoms. On the other hand, the charge on one mole of electrons is 96 485 C. This quantity is called the faraday (F).

The energies involved within individual atoms are likewise extremely small on the scale of the joule. Consequently it is convenient, in this context, to use the unit called the electron volt (eV), which is the quantity of energy required to raise a charge of one electron through a potential difference of 1 volt. Clearly, 1 eV $= 1.602 \times 10^{-19}$ J.

Another non-SI unit of energy is the calorie (cal). It was originally intended to be the energy required to raise the temperature of 1 g of water by 1 K, but this was found to be an imprecise definition. Currently, 1 cal is defined as 4.184 J, and it is widely used in the USA. In regard to nutrition, a unit frequently used is the 'large calorie' (Cal), meaning 10^3 cal or 1 kcal.

Glossary

It may be helpful to have a list of some technical terms used here, together with a brief explanation. Mostly, these will fall short of being definitions.

AC Alternating current. For example, mains electricity has alternating polarity.

aliquot A sample of a predetermined volume.

alkali metal The elements Li, Na, etc. of Group I.

alkaline earths The elements Be, Mg, etc. of Group II.

alkane The compounds of carbon and hydrogen of molecular formula, C_nH_{2n+2}, such as methane, ethane, propane, etc.

alkene The compounds of carbon and hydrogen of molecular formula, C_nH_{2n}, and containing one C=C double bond.

alkyl group A radical obtained (such as the ethyl radical) C_2H_5, when one H atom has been removed from an alkane.

alkyne The compounds of carbon and hydrogen of molecular formula, C_nH_{2n-2}, and containing one C≡C triple bond.

amino-acid A molecule which contains both a carboxylic acid group and an amine group. In the amino-acids of biological importance, both are attached to the same C atom. These are the building blocks of proteins, peptides and enzymes.

aromatic Pertaining to compounds which behave like benzene.

atom The smallest entity of an element, consisting of an atomic nucleus and the appropriate number of extranuclear electrons so that the atom is electrically neutral.

Avogadro's constant The number of molecules in one mole. Defined as the number of atoms of ^{12}C in 0.012 kg of ^{12}C.

axis of symmetry An axis, like the centre line of a wooden rail turned on a lathe, such that from any point, moving a certain distance perpendicular to the axis takes one to the same environment.

basis orbital An atomic orbital which serves as a component of a more sophisticated orbital.

bicyclic A molecule, such as naphthalene, which contains two rings.

buffer solution A solution of a definite pH whose pH value will be altered only marginally by the addition of a small amount of an acid or of a base.

burette A piece of laboratory equipment to facilitate the accurate measurement of any specified volume of solution.

carbohydrate A naturally occurring compound made up of large molecules, containing C, H and O, usually of a formula, $C_nH_{2m}O_m$, where n is slightly larger than m. These molecules are made up of residues of monosaccharides (such as glucose or fructose, both $C_6H_{12}O_6$) linked together by covalent bonds through an O atom.

centrifugal force The force that is required to be applied towards the centre of a circle in order to constrain an object to travel at constant speed around the circumference of the circle.

compound A substance which is made up of the atoms of more than one element. For example, water, H_2O.

coulombic forces The forces operative between electrically charged bodies. If charged in the same sense, the forces are repulsive; if in opposite senses, they are attractive.

cryoscopic Pertaining to the solidification of a liquid.

cyclic A molecule, such as cyclohexane, in which a ring exists because of the interatomic bonds.

DC Direct current, such as one would draw from a battery.

degenerate Electronic states of equal energy.

diamagnetic A substance with a negative magnetic susceptibility, which will tend to move out of an applied magnetic field. Most substances are diamagnetic.

ebullioscopic Pertaining to the vaporisation of a liquid.

elastic collision A collision with no loss of kinetic energy.

electrode A piece of metal, within a gas or a solution, to which a chosen voltage may be applied: if positively charged then it is an **anode**; if negative, a **cathode**.

electrolyte A substance which can dissociate in solution to give ions.

electromagnetic radiation Radiation, such as light or radio waves, involving mutually perpendicular electrical and magnetic fields.

electron A subatomic particle carrying a negative charge.

electrophile A reagent which attacks a region of excess negative charge within a molecule.

element A substance which cannot be broken down into different substances.

empirical equation An equation which is not based on a theoretical rationale but which describes observed behaviour.

end-point The stage in a titration at which the amounts of the two reacting substances conform to their ratio in the balanced equation for the reaction.

enzyme A molecule which acts as a catalyst within a biological system.

fat A naturally occurring compound consisting of three carboxylic acid residues covalently bonded in a specific way to a molecule of glycerol.

free radical A molecular entity in which the normal valency of one atom is not satisfied. For example, CH_3, the methyl radical, in which C is tervalent.

galvanic cell A cell which can act as a source of electric current.

glucose A particular type of sugar molecule, of molecular formula $C_6H_{12}O_6$, which serves as a building block for large molecules of carbohydrates.

glycogen A carbohydrate, built up in the muscles and the liver from glucose units.

halogen An element, like fluorine, belonging to Group VII with an electronic configuration ending in ns^5.

homologue A compound which is related to another by having one or several $-CH_2-$ units inserted.

hybrid orbital An orbital which is composed of a combination of primary atomic orbitals, thus possessing mixed character.

hydrolysis A reaction of a compound in which a bond is broken and 'the elements of water' are added to the two parts of the original molecule. For example, an alkyl chloride, RCl, yields the corresponding alkanol, ROH, along with HCl.

ideal solution A solution which obeys Raoult's Law.

ion An atom or a molecule to which an electron has been added or from which an electron has been removed. A positively charged ion is called a **cation**: a negatively charged ion is called an **anion**.

isoelectronic Having the same number of electrons, e.g. N_2, CO and CN^-.

isotope A nuclide (species of nucleus) with a specified atomic number or with the same atomic number as an identified nuclide. The term is sometimes applied to atoms.

kinetic energy The energy a body possesses by virtue of its motion.

latent heat The heat which a substance can absorb at constant temperature while it undergoes a change of phase.

mole That amount of any substance which contains a number of molecules equal to Avogadro's constant.

molecular orbital A concept used to interpret covalent interatomic bonds. The three-dimensional disposition of electron density around two or more nuclei.

molecule An assembly of atoms in which the combining power of every component atom is satisfied. For example, Ne, O_2 or CF_4.

neutralisation The process of chemical reaction between an acid and a base.

neutron An uncharged component of the atomic nucleus.

Newtonian mechanics The principles of physics summarised in Newton's (three) Laws of Motion.

noble gases The chemically inert elements of Group O.

nucleon An umbrella term for protons and neutrons, the component particles of the atomic nucleus.

nucleophile A chemical reagent which seeks to react with a part of a molecule or ion with excess positive charge.

Ohm's Law $V/I = R$. In the conduction of electricity through a body, the ratio of the applied voltage (V) to the current (I) is a constant, called the resistance (R).

orbit Like the track followed by a planet round the sun: the centrifugal force is matched by the force of attraction (in that case, gravitational) between the two bodies.

orbital The name coined, in the light of the Quantum theory, to describe the disposition of an electron in an atom or a molecule.

oxidation The reaction of a substance with oxygen, or any reaction which removes electrons from an atom.

paramagnetic A substance with a positive magnetic susceptibility, which will tend to move into a magnetic field. Most paramagnetic substances possess one or more unpaired electrons.

partial pressure The contribution of each component gas in a mixture of gases to the total pressure.

peptide A naturally occurring compound containing residues of amino acids, covalently bonded together via the linkage, –CO–NH–, where the –CO– comes from the acidic group of one residue and –NH– from the amine group of another.

Periodic Table A listing of the chemical elements in the manner initially achieved by Mendeleev, emphasising the periodicity that they demonstrate.

permittivity (relative) The attractive (or repulsive) forces between two unlike (or like) electric charges at a fixed separation depend on the medium between them. Taking as par the forces when the charges are in a vacuum, in the presence of any substance the force is diminished by a factor which is called the relative permittivity.

Planck's constant The fundamental constant by which the frequency of electromagnetic radiation needs to be multiplied to obtain the energy of a quantum.

protein A large peptide.

quantum The smallest possible amount of electromagnetic radiation.

reduction The reverse of **oxidation**: the reaction of a substance with hydrogen or any reaction which adds electrons to an atom.

relative permittivity The attractive (or repulsive) forces between two unlike (or like) electric charges at a fixed separation depend on the medium between them. Taking as par the forces when the charges are in a vacuum, in the presence of any substance the force is diminished by a factor which is called the relative permittivity.

reversible In regard to reaction mechanisms, a reversible step is one whose reverse cannot be disregarded. In thermodynamics, it means that a process is carried out infinitely slowly by a path (however hypothetical) which could allow one, on making an infinitesimal change to an intensive parameter, to go from the final to the initial state.

solute A substance which may be dissolved in a liquid, which plays the role of a **solvent**.

solvation The orienting of solvent molecules around a solute molecule (and, even more, around solute ions) so as to decrease the Gibbs energy of the system.

spherical symmetry If one goes a distance x from the centre in any direction, the value of the parameter is the same.

stoichiometric equation A balanced equation indicating the quantitative ratios involved between the molecules in a chemical reaction.

sublimation The direct conversion of a solid into a vapour.

substitution The replacement of one atom or radical within a molecule by another. For example, forming chloromethane (CH_3Cl) from methane (CH_4).

substrate A general term for a compound whose reactions are catalysed by an enzyme.

tautomerism The phenomenon of a compound having a constant molecular formula, but more than one molecular structure, which may interconvert.

unsaturated As applied to a hydrocarbon, this denotes the presence of one or more double or triple bonds between carbon atoms.

valency The capacity of atoms to form chemical bonds with other atoms: the number of chemical bonds which an atom of the element is able to form.

vitamin A compound whose ingestion is essential for good health, but which is required only in very small amounts.

Fundamental physical constants

Acceleration due to gravity, $g = 9.81$ m s^{-2}
Velocity of light *in vacuo*, $c = 2.998 \times 10^8$ m s^{-1}
Planck's constant, $h = 6.626 \times 10^{-34}$ J s
Boltzmann's constant, $k_B = 1.381 \times 10^{-23}$ J K^{-1}
Avogadro's constant, $N_A = 6.022 \times 10^{23}$ mol^{-1}
Gas constant, $R = 8.314$ J K^{-1} mol^{-1}
Electronic charge, $e = 1.602 \times 10^{-19}$ C
Faraday's constant, $F = 9.6485 \times 10^4$ C mol^{-1}
Atomic mass unit, a.m.u. $= 1.661 \times 10^{-27}$ kg
Mass of the electron, $m_e = 9.11 \times 10^{-31}$ kg

Answers to the problems

Chapter 1

1.1   (i) ^{102}Rn (ii) ^{60}Ni.

1.2   1.88 MeV. (Note that one does not need the mass of a β^- particle, because the figures given are atomic masses, rather than masses of the nuclei.)

1.3   0.054.

Chapter 2

2.1 $\nu = R_{\mathrm{H}} \left(\dfrac{1}{3^2} - \dfrac{1}{\mathrm{N}^2} \right).$ 1875, 1282 and 1093 nm.

2.2 5.75×10^{-3} Å.

2.3   At $r = 0$, a_0, $2a_0$, $3a_0$ and $4a_0$, $\psi/(1/\pi a_0{}^3)^{1/2}$ is 1.0, 0.368, 0.135, 0.050 and 0.018, $\psi^2/(1/\pi a_0^3)$ is 1.0, 0.135, 0.018, 0.0025 and 0.0003 and $4\pi r^2 \psi^2/a_0{}^{-1}$ is 0, 0.54, 0.29, 0.089 and 0.021.

Chapter 3

3.1   (a) sp^3, sp^2; (b) sp^3, sp, sp, sp^3, sp^3; 4 atoms collinear; (c) sp^2, sp^2, sp^3.

3.2   (a) diamagnetic; (b) paramagnetic.

Chapter 4

4.1   (a) 0.40 eV; (b) 2.25 eV; (c) 3.7×10^{-7} eV.

4.2   4119 cm^{-1}; 0.256 eV.

4.3   Vibrational.

4.4 $A = 0.833$; $\epsilon = 2082$ dm^3 mol^{-1} cm^{-1}.

4.5   Benzaldehyde.

4.6 1.77×10^{-4} mol dm^{-3}.

Chapter 5

5.1 6.6×10^{-8} moles.
5.2 1524 m s^{-1}.
5.3 5.67 m.

Chapter 6

6.1 15049 kJ.
6.2 -353.4 kJ mol^{-1}.
6.3 -89.7 kJ mol^{-1}.
6.4 1569.6 kJ; 403.3 g; 7.28°C.
6.5 -196.2 kJ mol^{-1}.
6.6 301.8 kJ mol^{-1}.
6.7 325.4 kJ mol^{-1}.
6.8 $+ 10.3$ kJ mol^{-1}; $+ 5.6$ kJ mol^{-1}.
6.9 2-propanol (112.5 J K^{-1} mol^{-1}) has a greater ΔS_{vap} than argon (71.8) or *n*-butane (81.6 J K^{-1} mol^{-1}).
6.10 $+210.6$ J K^{-1}.

Chapter 7

7.1 180 kJ mol^{-1}; 1442°C.
7.2 (a) 26.6 kJ mol^{-1}; (b) 76.4°C; (c) 0.011 atm.
7.3 (a) P $=$ 252 torr; y_B $=$ 0.53. (b) P $=$ 254 torr; y_B $=$ 0.41.
7.4 x_{PhCl} $=$ 0.75; y_{PhCl} $=$ 0.85.
7.5 Negative deviations. Strong forces of attraction between the two components.
7.6 0.027 mol dm^{-3}; less: in gas phase, 0.042 mol dm^{-3}.
7.7 0.0083.
7.8 260 g mol^{-1}.
7.9 (a) 1.86×10^{-3} degrees (b) 0.51×10^{-3} degrees (c) 0.00043 torr (d) 2477 Pa $\equiv$ 18.6 torr.
7.10 151 g mol^{-1}.
7.11 220 g mol^{-1}. The acid dimerises.

Chapter 8

8.1 129.7; -20.2 kJ mol^{-1}.
8.2 6.8×10^{12} bar^{-1}; 1.1×10^{-10} torr.
8.3 -103.9 kJ mol^{-1}; -227 J K^{-1} mol^{-1}; 1.7×10^{6}, more favourable.
8.4 1.4×10^{5} dm^{3} mol^{-1}.
8.5 2.9×10^{-4} mol dm^{-3}.

Chapter 9

9.1 5.9×10^{-4}.
9.2 0.115.
9.3 Add 14.1 cm^3 of NaOH solution to 20.0 cm^3 of the acid solution.
9.4 Add 15.1 cm^3 of sodium propanoate solution to 20.0 cm^3 of the acid solution.
9.5 Add 33.4 cm^3 of NaOH to 20.0 cm^3 of the oxalic acid solution.
9.6 Add 8.35 cm^3 of the HCl solution to 20.0 cm^3 of the ammonia solution.
9.7 ratio = 0.05.
9.8 3.66.
9.9 Thymol blue.
9.10 5.65.
9.11 No.

Chapter 10

10.1 -1.14 V.
10.2 -708 kJ mol^{-1}.
10.3 Ag, Hg, Cu.
10.4 (a) Yes; (b) $Br_2 + 3I^- = 2Br^- + I_3^-$;   (c) Yes;   (d) 6×10^7.
10.5 87 mV.

Chapter 11

11.1 $p = 1$; $q = 1$; $k = 0.12$ dm^3 mol^{-1} s^{-1}.
11.2 $k = 7.2 \times 10^{-3}$ s^{-1}; $t_{\frac{1}{2}} = 96$ s.
11.3 1.63×10^{-3} min^{-1} = 2.7×10^{-5} s^{-1}.
11.4 7.6×10^{-3} dm^3 mol^{-1} s^{-1}.
11.5 (a) increase by a factor of 3; (b) 3.6×10^{-3} mol dm^{-3} s^{-1}; (c) decrease by a factor of $2^3 = 8$; (d) third order rate constant: dm^6 mol^{-2} s^{-1}.
11.6 116.2 kJ mol^{-1}; 9.4×10^{15} s^{-1}.

Chapter 12

12.1 $k' = k_1 + k_2 [OH^-]$

12.2 Rate $= \dfrac{k_1 k_2 [Ce^{4+}][Ag^+][Tl^+]}{k_{-1}[Ce^{3+}] + k_2[Tl^+]}$

12.3 $k_f = 6.9 \times 10^{-3}$ min^{-1}; $k_r = 2.6 \times 10^{-3}$ min^{-1}.
12.4 $v_{max} = 23.3 \times 10^{-8}$ mol dm^{-3} s^{-1}; $K_m = 4.7 \times 10^{-3}$.

Index